Hypothalamic-Pituitary Development

Endocrine Development

Vol. 4

Series Editor

Martin O. Savage London

KARGER

Hypothalamic-Pituitary Development

Genetic and Clinical Aspects

Volume Editors

Raphaël Rappaport *Paris*
Serge Amselem *Créteil*

34 figures, 2 in color, 22 tables, 2001

KARGER

Basel · Freiburg · Paris · London · New York ·
New Delhi · Bangkok · Singapore · Tokyo · Sydney

Raphaël Rappaport
Hôpital Necker-Enfants Malades
Paris (France)

Serge Amselem
Institut National de la Santé et de la Recherche Médicale
Hôpital Henri-Mondor
Créteil (France)

Library of Congress Cataloging-in-Publication Data

Hypothalamic-pituitary development : genetic and clinical aspects / volume editors, Raphaël Rappaport, Serge Amselem.
p. ; cm. – (Endocrine development, ISSN 1421–7082; vol. 4)
Includes bibliographical references and index.
ISBN 3805572395 (hard cover : alk. paper)
1. Pituitary gland–Pathophysiology. 2. Hypothalamus–Pathophysiology.
3. Endocrinology, Developmental. 4. Pituitary gland–Growth. 5. Pituitary gland–Growth.
I. Rappaport R. (Raphaël) II. Anselem, Serge. III. Series.
[DNLM: 1. Hypothalamus–growth & development.
2. Hypothalamic Diseases–genetics. 3. Pituitary Gland–growth & development. WL 312 H99863 2001]
RC658.H97 2001
616.4′7–dc21 2001029954

Bibliographic Indices. This publication is listed in bibliographic services, including Current Contents® and Index Medicus

Drug Dosage. The authors and the publisher have exerted every effort to ensure that drug selection and dosage set forth in this text are in accord with current recommendations and practice at the time of publication. However, in view of ongoing research, changes in government regulations, and the constant flow of information relating to drug therapy and drug reactions, the reader is urged to check the package insert for each drug for any change in indications and dosage and for added warnings and precautions. This is particularly important when the recommended agent is a new and/or infrequently employed drug.

www.karger.com
Printed in Switzerland on acid-free paper by Reinhardt Druck, Basel
ISSN 1421–7082
ISBN 3–8055–7239–5

Contents

Foreword

The fourth volume in the series *Endocrine Development* continues the pattern established in earlier volumes of combining discussions on fundamental scientific and clinical advances. Professor Raphaël Rappaport and Professor Serge Amselem, themselves experts in clinical and molecular endocrinology respectively, have assembled an impressive list of world authorities to contribute to this volume. Hypothalamic-pituitary development is at the origin of many essential endocrine systems and defects in this developmental process cause a wide spectrum of clinical defects relevant to both paediatric and adult endocrinology.

The genetic basis of many of these defects has been established in the past 20 years. The book includes descriptions of normal and abnormal neuroendocrine development and function. Anatomical and genetic derangements are described and linked to their clinical syndromes.

Overall, this book continues the long-term aim of the series, which is to explain, instruct and stimulate scientists and clinicians to increase their knowl edge of important and clinically relevant endocrine disorders.

London, March 2001 *Martin O. Savage*

Preface

Over recent years, dramatic progress has been made in understanding pituitary insufficiency in the human by the combination of experimental studies, molecular genetics and clinical research. The description of pituitary gland development has opened a new and exciting field of research and provided clues to the ill-defined group of idiopathic isolated growth hormone deficiency or multiple pituitary hormone deficiency.

The unique opportunity to provide an updated and prospective view of this field was given to us when it was proposed by Professor M. Savage to edit a book covering this rapidly expanding knowledge. All authors have accepted the challenge to integrate complex knowledge from various scientific and medical fields providing updated and didactic information with new approaches in developmental endocrinology.

Anterior and hypothalamic organogenesis is described in animal models with particular reference to the human. Whenever possible, emphasis is put on the combination of pituitary defects with a multitude of facial and cranial abnormalities, some of which are part of well-described syndromes like septo-optic dysplasia. An effort is made to correlate genotypic and phenotypic data in order to support innovative practice in pediatrics and endocrinology.

Fine imaging of the brain and the pituitary, using magnetic resonance, contributes to define new organic pituitary defects reducing the component of 'idiopathic' conditions with growth hormone deficiency. The pituitary stalk interruption syndrome with its various presentations remains unexplained although suggesting strong interactions between hypothalamus and anterior pituitary during embryogenesis.

Exciting clinical data are also reported, indicating the role of a number of homeodomain transcription factors regulating pituitary cell specification and thereby various combinations of hormonal defects. Several months or years after the initial reports, some of which were made by contributors to this book, follow-up of patients and reports of new cases have added valuable new information for the clinician and the genetist. In addition, various aspects of posterior pituitary development and dysfunction are described with particular emphasis on recent imaging data.

It was the intention of the editors and the contributors to help clinicians and to stimulate their interest in unique models of genetic and developmental defects of the pituitary which bear ultimate consequences on metabolic control, growth and reproduction. It is our wish that this effort might have contributed to provide an easy entry to modern 'molecular medicine' as well as to 'developmental endocrinology'.

Raphaël Rappaport
Serge Amselem

Rappaport R, Amselem S (eds): Hypothalamic-Pituitary Development. Genetic and Clinical Aspects.
Endocr Dev. Basel, Karger, 2001, vol 4, pp 1–12

Genetic Regulation of Forebrain and Pituitary Development

Donna M. Martin[a], *Sally A. Camper*[b]

Departments of
[a] Pediatrics and
[b] Human Genetics, The University of Michigan, Ann Arbor, Mich., USA

Hypothalamic-Pituitary Function – An Overview

The mature pituitary gland is comprised of three lobes: anterior, intermediate and posterior (fig. 1). The anterior lobe contains five major cell types, defined by the hormones they produce: somatotropes (growth hormone, GH), lactotropes (prolactin, PRL), gonadotropes (luteinizing hormone, LH, and follicle-stimulating hormone, FSH), corticotropes (adrenocorticotropic hormone, ACTH), and thyrotropes (thyroid-stimulating hormone, TSH). In response to neural induction from the hypothalamus, hormonal release from anterior pituitary cells regulates body temperature, growth, sexual maturity, fertility, lactation and stress responses. The intermediate lobe of the pituitary produces pro-opiomelanocortin (POMC), the precursor protein to melanocyte-stimulating hormone (MSH) and endorphins. The posterior lobe (neurohypophysis) of the pituitary is of neuronal origin, and contains the synaptic terminals of neurons whose cell bodies reside in the hypothalamus and secretes oxytocin and vasopressin, hormones that regulate lactation and volume status, respectively.

The hypothalamus lies anterior to the developing pituitary, and contains several discrete brain nuclei with neural projections to multiple brain regions, including the median eminence and posterior pituitary. The median eminence, a portal vascular system that lies in close proximity to nerve terminals from the hypothalamus, carries hormones from these terminals to their target cell types in the anterior pituitary. Parvocellular neurons of the PVN project to the median eminence, where they release corticotropin-releasing hormone (CRH) and thyroid-releasing hormone (TRH) which travel to the anterior pituitary and regulate the pituitary-adrenal and pituitary-thyroid hormone axes, respectively. Neurons of the arcuate

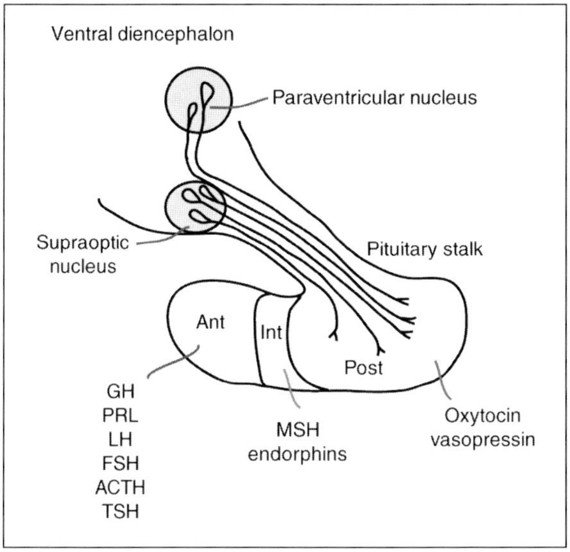

Fig. 1. Schematic of the pituitary gland and ventral diencephalon. Major hypothalamic nuclei (supraoptic nucleus and paraventricular nucleus) are located in the ventral diencephalon. Their axons project through the pituitary stalk to the posterior pituitary. The three lobes of the pituitary (anterior, posterior and intermediate) are shown, along with hormones secreted in each. GH = Growth hormone, PRL = prolactin, LH = luteinizing hormone, FSH = follicle-stimulating hormone, ACTH = adrenocorticotropic hormone, TSH = thyroid-stimulating hormone, MSH = melanocyte-stimulating hormone.

hypothalamic nucleus also project to the median eminence, where they secrete dopamine and somatostatin, regulators of PRL and GH secretion. Magnocellular neurons of the medially located PVN and SON of the hypothalamus project to the posterior pituitary, where they secrete oxytocin and vasopressin.

This chapter focuses on the developmental genetic mechanisms regulating normal hypothalamic-pituitary development, with an emphasis on known gene mutations in mice that in some cases have led to the identification of similar mutations in humans.

Pituitary and Hypothalamic Development

The anterior and intermediate lobes of the pituitary develop from Rathke's pouch, which is initially derived from the anterior neural ridge (ANR) adjacent to cells in the neural plate that contribute to ventral diencephalon [1]. Invagination of oral ectoderm and subsequent separation of Rathke's pouch from the oral ectoderm, along with extensive cell proliferation and differentiation, results in the final mature pituitary gland. Ablation of the central ANR in 3–4 somite stage chicks completely disrupts development of Rathke's pouch and results in pituitary aplasia [2]. Ectopic placement of ventral diencephalon from 2- to 4-day-old chick embryos into head ectoderm induces differentiation of anterior pituitary cell types, suggesting that contact between these two primordia are necessary for proper anterior pituitary development [3]. In vitro organ culture in chicks has confirmed this observation, and further demonstrated a crucial role for mesenchyme and notochord in outpouching of the oral ectoderm. Mouse ventral diencephalon also induces anterior pituitary differentiation in the chick, indicating conservation of early pathways between these two species [4].

The posterior lobe of the pituitary develops from the infundibulum, an evagination of the ventral diencephalon. Development of the ventral diencephalon is thought to follow the prosomeric theory of central nervous system development [5–7]. According to this theory, longitudinal and transverse domains (prosomeres) along the developing rostral neural tube serve to define functional units that can be identified by morphological and gene expression analyses. The neuronal organization of the diencephalon is complex. Unlike other brain regions such as the cerebral cortex, where newly formed cells migrate radially from the ventricular layer to form a laminar structure, diencephalic nuclei develop through highly specific migration and differentiation of neurons and glia into their final functional locations. Studies using retroviral markers in chick demonstrate that single clonal neural progenitors arise from the ventricular zone of the third ventricle and give rise to neurons, glia and radial glia [8]. These related cells often reside within different diencephalic nuclei that span the midline and are physically distant from each other. Migration of neural progenitors and their offspring to their final nuclear locations appears to parallel the pattern of radial glia; however, most neural progenitor cells in the developing ventral diencephalon are not dedicated to producing all the cells for a particular hypothalamic nucleus or group of nuclei.

Genetic Control of Hypothalamic-Pituitary Development

Over the past 20 years, our understanding of the genetic regulation of hypothalamic-pituitary development has increased exponentially through the study of human mutations, animal models, and tissue culture systems. Our laboratory and others have focused on the study of spontaneous and artificially induced mutations in mice with pituitary dysfunction [9–15]. Mutations in at least 27 genes, including transcription factors, soluble proteins, hormones, cell surface and nuclear receptors disrupt pituitary and/or hypothalamic development (table 1),

Table 1. Genes involved in hypothalamic-pituitary development

Gene	Protein	Organism/mutation	Phenotype	Ref.
Transcription factors				
Brn2 or *POU3F2*	Class III POU HD TF	Mouse KO	Loss of SON, PVN and posterior pituitary	22
Dax1 or *Ahch*	Transcription factor	Mouse KO	Adrenal hypoplasia congenita and hypogonadotropic hypogonadism	25
Egr1	Zinc finger TF	Mouse KO	Reduced growth, female infertility, reduced somatotropes	26, 27
Ftzf1	SF1, steroidogenic factor 1 (fushi tarizu 1 factor homologue)	Mouse KO	Lethal; gonadotrope-deficient; adrenal and gonad-deficient, defects of the VMH	23, 28
Gata2	Zinc finger TF	Mouse KO, dominant negative TG	Lethal due to anemia; role in thyrotrope and gonadotrope fate determination	29, 30
Gli1	Zinc finger TF	Mouse KO	Contributes to pituitary and ventral diencephalon abnormalities in presence of *Gli2* or *Gli3* mutations	31
Gli2	Zinc finger TF	Mouse KO	Variable absence of the pituitary, abnormal midline ventral diencephalon	31
Hesx1 or *Rpx*	Paired-like HD TF	Mouse KO	Pituitary hypoplasia, microphthalmia, agenesis of corpus callosum and septum pellucidum	32
Isl1	LIM HD TF	Mouse KO	Lethal; implicated in corticotrope and thyrotrope patterning	33
Lhx3	LIM HD protein-3 TF	Mouse KO	Lethal; loss of thyrotropes, somatotropes, lactotropes, gonadotropes	34
Lhx4 or *Gsh1*	LIM HD protein-4 TF	Mouse KO	Dwarf; reduced somatotropes and lactotropes; altered GHRH expression in arcuate nucleus of hypothalamus	35, 36
Otx1	Bicoid-like HD TF	Mouse KO	Transient hypogonadism and dwarfism	37
Pax6	Paired-like HD TF	Mouse KO	Ventralization of dorsal-ventral pituitary axis	38
Prop1	Paired-like HD TF	Ames dwarf (*df*, $Prop1^{df}$)	Loss of most somatotropes, lactotropes and thyrotropes; reduced gonadotropes	39, 40

Table 1 (continued)

Gene	Protein	Organism/mutation	Phenotype	Ref.
Pit1	GHF1, pituitary-specific POU HD TF	Snell dwarf (*dw, Pit1^{dw}*); Jackson dwarf (*dw^{J}, Pit1$^{dw\text{-}J}$*)	Loss of somatotropes, lactotropes and thyrotropes; increased gonadotropes	9, 41
Pitx1	Bicoid-like HD TF	Mouse KO	Decreased LH, FSH, TSH, increased ACTH	42, 43
Pitx2	Bicoid-like HD TF	Mouse KO	Lethal; pituitary hypoplasia, severe defects of the ventral abdominal wall, heart eyes and teeth	44–47
Titf1	Thyroid TF	Mouse KO	Loss of anterior and intermediate lobes	48
Soluble factors/hormones				
Gnrh	GNRH	*hpg* mice	Reduced gonadotrope number	49
α-GSU	Pituitary glycoprotein α subunit common to FSH, LH, TSH	Mouse KO	Dwarfism, hypogonadism, hypothyroidism	50
Wnt4	Wingless-like protein type 4	Mouse KO	Hypoplastic pituitary; decreased gonadotropes, thyrotropes and somatotropes	51
BMP4	Bone morphogenetic protein-4	Mouse KO	Lethal; absence of thickened ectodermal pouch placode	52
Shh	Sonic hedgehog protein	Mouse KO	Absence of pituitary; dysmorphic ventral diencephalon	53
Receptors				
ActRCII	Activin receptor	Mouse KO	Loss of FSH in gonadotropes	54
Dat1	Dopamine transporter 1	Mouse KO	Reduced somatotrope and lactotrope number; dwarf; GHRH deficiency	55
Drd2	Dopamine receptor 2	Mouse KO	Lactotrope hyperplasia and tumor formation	56
Ghrhr	GHRH receptor	Little (*lit, Ghrhr^{lit}*) mouse	Reduced somatotrope number	57, 58

TF = Transcription factor, HD = homeodomain, KO = knockout, FSH = follicle-stimulating hormone, SON = supraoptic nucleus, PVN = paraventricular nucleus, VMH = ventromedial nucleus, GHRH = growth hormone-releasing hormone, GNRH = gonadotropin releasing hormone. Only those mutations in mice demonstrated as having a direct role in hypothalamic-pituitary development are included. Please see original literature for details.

Table 2. Human mutations affecting hypothalamic-pituitary development

Gene	Protein	Condition/phenotype	Ref.
DAX1 or *ACH*	Transcription factor	Adrenal hypoplasia congenita and hypogonadotropic hypogonadism	59, 60
GHRHR	Growth hormone-releasing hormone receptor	Isolated GH deficiency	61
GLI3	Zinc finger TF	Pallister-Hall syndrome	62
GNAS1	Ghrhr-associated Gas	Acromegaly; pituitary hyperplasia, GH-secreting tumors	63
GNRHR	Gonadotropin-releasing hormone receptor	Hypogonadotropic hypogonadism; FSH/LH reduction; pubertal delay	64
HESX1 or *RPX*	HD TF	Septo-optic dysplasia/ congenital pituitary hypoplasia	32, 65
LHX3	LIM HD TF	CPHD	66
PROP1	Paired-like HD transcription factor-1	CPHD; loss of GH, PRL and TSH, reduction in LG and FSH	67, 68
POU1F1 or *PIT1*	Pituitary-specific transcription factor 1 (PIT1)	Panhypopituitarism/loss of GH and PRL; variable TSH loss	69

Only those mutations demonstrated as having a direct role in hypothalamic-pituitary development are included. Please see original literature for details.

some of which have corresponding human mutations (table 2). We have also employed the differential display method to identify genes whose expression pattern is dynamic during the period of highest pituitary proliferation, between embryonic days 12.5 and 14.5 in the mouse [16]. This method has led to the identification of 83 transcripts, 40% of which are novel. Approaches such as this are likely to yield candidate genes for hypothalamic and pituitary developmental regulation and disease.

Early Genetic Interactions in the Pituitary and Ventral Diencephalon

The ventral diencephalon, in close proximity to Rathke's pouch, exerts inductive influences on normal pituitary development. In mice mutant for *Titf1*, a gene expressed in developing brain but not pituitary, complete pituitary aplasia is observed [17]. Bone morphogenetic protein 4 (BMP4) is produced in the embryonic day 8.5 mouse ventral diencephalon and is required for normal formation of Rathke's pouch [17–19]. BMP4 is also necessary for maintenance of *Isl1*, a transcription factor expressed in ventral anterior pituitary cells that probably later become the rostral tip thyrotropes. The ventral diencephalon also produces fibroblast growth factor 8 (FGF8) a factor necessary for activating *Lhx3*, a LIM-family homeodomain gene. A critical period appears to exist during which molecules like BMP4 and FGF8 from the ventral diencephalon are necessary for inducing pouch formation. In the mouse, this critical period ends when the pouch separates from the oral ectoderm (e12–13), after which the rudimentary pouch can be explanted in culture and successfully stimulated to form all the various cell types [20, 21].

The earliest transcription factors expressed in the developing pouch are LHX3, LHX4, PITX1 and PITX2, each of which is involved in expansion of the committed pouch early in gestation. *RPX*, essential for normal pouch formation, activates *PROP1*, the gene mutated in *Ames dwarf* mice and patients with combined pituitary hormone deficiency. *Prop1* expression instructs differentiation of the gonadotrope, thyrotrope, lactotrope and somatotrope cell lineages, activates *Pit1* and suppresses *Rpx*. The corticotrope lineage develops independently of PROP1 and downstream targets, and is instead instructed by *Rpx* expression. Other transcription factor genes, including *Nr5a1* (encodes SF1), *Nr4a1* or *Nurr77, NeuroD1, Gata2, Egr1, Otx1, Zn15* and *ER* (encodes the estrogen receptor) act at more specialized locations in the genetic hierarchy, instructing differentiation or proliferation of individual pituitary cell types.

The mechanisms by which highly specialized individual nuclei of the hypothalamus (i.e. paraventricular, supraoptic and arcuate nuclei) develop from their ancestral neural progenitors in the ventricular zone of the third ventricle are poorly understood. This process is surely dependent on genes that regulate patterning in the forebrain, which are now emerging as candidates for hypothalamic cell fate determination. Examination of mice with loss of Brn-4, a class III POU domain transcription factor also known as POU3F2, revealed loss of PVN and SON, with inability of neurons in these nuclei to differentiate and activate downstream genes crucial for their survival [22]. Similarly, mice with mutations in *Nr5a1* (or *Ftzf1*), the gene encoding the nuclear receptor SF-1, lack the ventromedial hypothalamic nucleus [23]. These observations demonstrate that loss of single genes can ablate specific hypothalamic nuclei and raise the possibility that similar defects exist in humans but are currently undetected given that hypothalamic nuclei are below the available degree of neuroimaging resolution. Many transcription factors affect multiple organ systems; one way to sort out their effects on the development of individual organs is to use inducible gene ablation using organ-specific and temporally-regulated expression. For example, the roles of SF1 in brain, pituitary, adrenal and gonads are currently being examined using Cre-loxP technology [24].

Concluding Remarks

The precise temporal and spatial expression of pituitary and hypothalamic cellular phenotypes has served as an excellent model for understanding the interactions between genes and environment during development. Through the use of spontaneous and genetically engineered mouse mutations, we now have a working knowledge of the early developmental processes by which these various cell types develop. Moreover, several human mutations are now known to disrupt normal hypothalamic-pituitary development, and many more are sure to be identified. We are entering an era of tissue and cell-type specific developmental genetic analyses that will allow us to determine precisely the time and place where hypothalamic and pituitary defects in mice occur. This knowledge will increase our ability to characterize known mutations in people with hypothalamic-pituitary dysfunction more completely, will enable more extensive human mutation screening in individuals who currently lack a molecular diagnosis, and improve potential therapies for hypothalamic-pituitary tumors.

Acknowledgements

We thank Frank Probst for assistance with figures and critical manuscript review, and Merle Rozensweig for helping with references.

References

1 Couly G, Le Douarin NM: The fate map of the cephalic neural primordium at the presomitic to the 3-somite stage in the avian embryo. Development 1988;103(suppl):101–113.

2 El Amraoui A, Dubois PM: Experimental evidence for the early commitment of the presumptive adenohypophysis. Neuroendocrinology 1993;58:609–615.

3 Fedtsova NG, Barabanov VM: The distribution of competence for adenohypophysis development in the ectoderm of chick embryos. Ontogenez 1990;21:254–260.

4 Gleiberman AS, Fedtsova NG, Rosenfeld MG: Tissue interactions in the induction of anterior pituitary: Role of the ventral diencephalon, mesenchyme, and notochord. Dev Biol 1999;213:340–353.

5 Puelles L, Rubenstein JL: Expression patterns of homeobox and other putative regulatory genes in the embryonic mouse forebrain suggest a neuromeric organization. Trends Neurosci 1993;16:472–479.

6 Rubenstein JL, Martinez S, Shimamura K, Puelles L: The embryonic vertebrate forebrain: The prosomeric model. Science 1994;266:578–580.

7 Figdor MC, Stern CD: Segmental organization of embryonic diencephalon. Nature 1993;363:630–634.

8 Golden JA, Cepko CL: Clones in the chick diencephalon contain multiple cell types and siblings are widely dispersed. Development 1996;22:65–78.

9 Camper SA, Saunders TL, Katz RW, Reeves RH: The Pit-1 transcription factor gene is a candidate for the murine Snell dwarf mutation. Genomics 1990;8:586–590.

10 Lossie AC, Gordon DF, Camper SA: Localization of thyrotropin-releasing hormone receptor and thyrotroph embryonic factor on mouse chromosome 15. Mamm Genome 1993;4:621–623.

11 Knapp LT, Keegan CE, Seasholtz AF, Camper SA: Corticotropin-releasing hormone (Crh) maps to mouse chromosome 3. Mamm Genome 1993;4:615–617.

12 Roller ML, Camper SA: Localization of the thyrotropin-releasing hormone gene, Trh, on mouse chromosome 6. Mamm Genome 1995;6:443–444.

13 Gage PJ, Camper SA: Pituitary homeobox 2, a novel member of the bicoid-related family of homeobox genes, is a potential regulator of anterior structure formation. Hum Mol Genet 1997;6:457–464.

14 Lesh JS, Burrows HL, Seasholtz AF, Camper SA: Mapping of the mouse corticotropin-releasing hormone receptor 2 gene (Crhr2) to chromosome 6. Mamm Genome 1997;8:944–945.

15 Watkins-Chow DE, Camper SA: How many homeobox genes does it take to make a pituitary gland? Trends Genet 1998;14:284–290.

16 Douglas KR, Camper SA: Partial transcriptome of the developing pituitary gland. Genomics 2000; 70:335–346.

17 Takuma N, Sheng HZ, Furuta Y, Ward JM, Sharma K, Hogan BL, Pfaff SL, Westphal H, Kimura S, Mahon KA: Formation of Rathke's pouch requires dual induction from the diencephalon. Development 1998;125:4835–4840.

18 Ericson J, Norlin S, Jessell TM, Edlund T: Integrated FGF and BMP signaling controls the progression of progenitor cell differentiation and the emergence of pattern in the embryonic anterior pituitary. Development 1998;125:1005–1015.

19 Treier M, Gleiberman AS, O'Connell SM, Szeto DP, McMahon JA, McMahon AP, Rosenfeld MG: Multistep signaling requirements for pituitary organogenesis in vivo. Genes Dev 1998;12:1691–1704.

20 Nemeskery A, Nemeth A, Setalo G, Vigh S, Halasz B: Cell differentiation of the fetal rat anterior pituitary in vitro. Cell Tissue Res 1976;170:263–273.

21 Dubois PM, Hemming FJ: Fetal development and regulation of pituitary cell types. J Electron Microsc Tech 1991;19:2–20.

22 Schonemann MD, Ryan AK, McEvilly RJ, O'Connell SM, Arias CA, Kalla KA, Li P, Sawchenko PE, Rosenfeld MG: Development and survival of the endocrine hypothalamus and posterior pituitary gland requires the neuronal POU domain factor Brn-2. Genes Dev 1995;9:3122–3135.

23 Ikeda Y, Luo X, Abbud R, Nilson JH, Parker KL: The nuclear receptor steroidogenic factor 1 is essential for the formation of the ventromedial hypothalamic nucleus. Mol Endocrinol 1995;9:478–486.

24 Zhao L, Bakke M, Krimkevich Y, Cushman LJ, Parlow AF, Camper SA, Parker KL: Steroidogenic factor 1 (SF1) is essential for pituitary gonadotrope function. Development 2001;128:147–154.

25 Yu F, Venzon DJ, Serrano J, Goebel SU, Doppman JL, Gibril F, Jensen RT: Prospective study of the clinical course, prognostic factors, causes of death, and survival in patients with long-standing Zollinger-Ellison syndrome. J Clin Oncol 1999;17:615–630.

26 Topilko P, Schneider-Maunoury S, Levi G, Trembleau A, Gourdji D, Driancourt MA, Rao CV, Charnay P: Multiple pituitary and ovarian defects in Krox-24 (NGFI-A, Egr-1)-targeted mice. Mol Endocrinol 1998;12:107–122.

27 Lee SL, Sadovsky Y, Swirnoff AH, Polish JA, Goda P, Gavrilina G, Milbrandt J: Luteinizing hormone deficiency and female infertility in mice lacking the transcription factor NGFI-A (Egr-1). Science 1996;273:1219–1221.

28 Ingraham HA, Lala DS, Ikeda Y, Luo X, Shen WH, Nachtigal MW, Abbud R, Nilson JH, Parker KL: The nuclear receptor steroidogenic factor 1 acts at multiple levels of the reproductive axis. Genes Dev 1994;8:2302–2312.

29 Tsai FY, Keller G, Kuo FC, Weiss M, Chen J, Rosenblatt M, Alt FW, Orkin SH: An early haematopoietic defect in mice lacking the transcription factor GATA-2. Nature 1994;371:221–226.

30 Dasen JS, O'Connell SM, Flynn SE, Treier M, Gleiberman AS, Szeto DP, Hooshmand F, Aggarwal AK, Rosenfeld MG: Reciprocal interactions of Pit1 and GATA2 mediate signaling gradient-induced determination of pituitary cell types. Cell 1999;97:587–598.

31 Park HL, Bai C, Platt KA, Matise MP, Beeghly A, Hui CC, Nakashima M, Joyner AL: Mouse Gli1 mutants are viable but have defects in SHH signaling in combination with a Gli2 mutation. Development 2000;127:1593–1605.

32 Dattani MT, Martinez-Barbera JP, Thomas PQ, Brickman JM, Gupta R, Martensson IL, Toresson H, Fox M, Wales JK, Hindmarsh PC, Krauss S, Beddington RS, Robinson IC: Mutations in the homeobox gene HESX1/Hesx1 associated with septo-optic dysplasia in human and mouse. Nat Genet 1998;19:125–133.

33 Pfaff SL, Mendelsohn M, Stewart CL, Edlund T, Jessell TM: Requirement for LIM homeobox gene Isl1 in motor neuron generation reveals a motor neuron-dependent step in interneuron differentiation. Cell 1996;84:309–320.

34 Sheng HZ, Zhadanov AB, Mosinger B, Fujii T, Bertuzzi S, Grinberg A, Lee EJ, Huang SP, Mahon KA, Westphal H: Specification of pituitary cell lineages by the LIM homeobox gene Lhx3. Science 1996;272:1004–1007.

35 Sheng HZ, Moriyama K, Yamashita T, Li H, Potter SS, Mahon KA, Westphal H: Multistep control of pituitary organogenesis. Science 1997;278:1809–1182.

36 Li H, Witte DP, Branford WW, Aronow BJ, Weinstein M, Kaur S, Wert S, Singh G, Schreiner CM, Whitsett JA et al: Gsh-4 encodes a LIM-type homeodomain, is expressed in the developing central nervous system and is required for early postnatal survival. EMBO J 1994;13:2876–2885.

37 Acampora D, Mazan S, Tuorto F, Avantaggiato V, Tremblay JJ, Lazzaro D, di Carlo A, Mariano A, Macchia PE, Corte G, Macchia V, Drouin J, Brulet P, Simeone A: Transient dwarfism and hypogonadism in mice lacking Otx1 reveal prepubescent stage-specific control of pituitary levels of GH, FSH and LH. Development 1998;125:1229–1239.

38 Kioussi C, O'Connell S, St-Onge L, Treier M, Gleiberman AS, Gruss P, Rosenfeld MG: Pax6 is essential for establishing ventral-dorsal cell boundaries in pituitary gland development. Proc Natl Acad Sci USA 1999;96:14378–14382.

39 Sornson MW, Wu W, Dasen JS, Flynn SE, Norman DJ, O'Connell SM, Gukovsky I, Carriere C, Ryan AK, Miller AP, Zuo L, Gleiberman AS, Andersen B, Beamer WG, Rosenfeld MG: Pituitary lineage determination by the Prophet of Pit-1 homeodomain factor defective in Ames dwarfism. Nature 1996;384:327–333.

40 Gage PJ, Brinkmeier ML, Scarlett LM, Knapp LT, Camper SA, Mahon KA: The Ames dwarf gene, df, is required early in pituitary ontogeny for the extinction of Rpx transcription and initiation of lineage-specific cell proliferation. Mol Endocrinol 1996;10:1570–1581.

41 Li S, Crenshaw EB, Rawson EJ, Simmons DM, Swanson LW, Rosenfeld MG: Dwarf locus mutants lacking three pituitary cell types result from mutations in the POU-domain gene pit-1. Nature 1990; 347:528–533.

42 Szeto DP, Rodriguez-Esteban C, Ryan AK, O'Connell SM, Liu F, Kioussi C, Gleiberman AS, Izpisua-Belmonte JC, Rosenfeld MG: Role of the bicoid-related homeodomain factor Pitx1 in specifying hindlimb morphogenesis and pituitary development. Genes Dev 1999;13:484–494.

43 Lanctot C, Gauthier Y, Drouin J: Pituitary homeobox 1 (Ptx1) is differentially expressed during pituitary development. Endocrinology 1999;140:1416–1422.

44 Kitamura K, Miura H, Miyagawa-Tomita S, Yanazawa M, Katoh-Fukui Y, Suzuki R, Ohuchi H, Suehiro A, Motegi Y, Nakahara Y, Kondo S, Yokoyama M: Mouse Pitx2 deficiency leads to anomalies of the ventral body wall, heart, extra- and periocular mesoderm and right pulmonary isomerism. Development 1999;126:5749–5758.

45 Gage PJ, Suh H, Camper SA: Dosage requirement of Pitx2 for development of multiple organs. Development 1999;126:4643–4651.
46 Lin CR, Kioussi C, O'Connell S, Briata P, Szeto D, Liu F, Izpisua-Belmonte MG, Rosenfeld MG: Pitx2 regulates lung asymmetry, cardiac positioning and pituitary and tooth morphogenesis. Nature 1999;401:279–282.
47 Lu MF, Pressman C, Dyer R, Johnson RL, Martin JF: Function of Rieger syndrome gene in left-right asymmetry and craniofacial development. Nature 1999;401:276–278.
48 Kimura S, Hara Y, Pineau T, Fernandez-Salguero P, Fox CH, Ward JM, Gonzalez FJ: The T/ebp null mouse: Thyroid-specific enhancer-binding protein is essential for the organogenesis of the thyroid, lung, ventral forebrain, and pituitary. Genes Dev 1996;10:60–69.
49 McDowell IF, Morris JF, Charlton HM: Characterization of the pituitary gonadotroph cells of hypogonadal (hpg) male mice: Comparison with normal mice. J Endocrinol 1982;95:321–330.
50 Kendall SK, Samuelson LC, SaundersTL, Wood RI, Camper SA: Targeted disruption of the pituitary glycoprotein hormone alpha-subunit produces hypogonadal and hypothyroid mice. Genes Dev 1995;9:2007–2019.
51 Ikeya M, Lee SM, Johnson JE, McMahon AP, Takada S: Wnt signalling required for expansion of neural crest and CNS progenitors. Nature 1997;389:966–970.
52 Winnier G, Blessing M, Labosky PA, Hogan BL: Bone morphogenetic protein-4 is required for mesoderm formation and patterning in the mouse. Genes Dev 1995;9:2105–2016.
53 St-Jacques B, Dassule HR, Karavanova I, Botchkarev VA, Li J, Danielian PS, McMahon JA, Lewis PM, Paus R, McMahon AP: Sonic hedgehog signaling is essential for hair development. Curr Biol 1998;8:1058–1068.
54 Matzuk MM, Kumar TR, Bradley A: Different phenotypes for mice deficient in either activins or activin receptor type II. Nature 1995;374:356–360.
55 Bosse R, Fumagalli F, Jaber M, Giros B, Gainetdinov RR, Wetsel WC, Missale C, Caron MG: Anterior pituitary hypoplasia and dwarfism in mice lacking the dopamine transporter. Neuron 1997;19:127–138.
56 Saiardi A, Bozzi Y, Baik JH, Borrelli E: Antiproliferative role of dopamine: loss of D2 receptors causes hormonal dysfunction and pituitary hyperplasia. Neuron 1997;19:115–126.
57 Mayo KE: A little lesson in growth regulation. Nat Genet 1996;12:8–9.
58 Lin SC, Lin CR, Gukovsky I, Lusis AJ, Sawchenko PE, Rosenfeld MG: Molecular basis of the little mouse phenotype and implications for cell type-specific growth. Nature 1993;364:208–213.
59 Muscatelli F, Strom TM, Walker AP, Zanaria E, Recan D, Meindl A, Bardoni B, Guioli S, Zehetner G, Rabl W et al: Mutations in the DAX-1 gene give rise to both X-linked adrenal hypoplasia congenita and hypogonadotropic hypogonadism. Nature 1994;372:672–676.
60 Zanaria E, Muscatelli F, Bardoni B, Strom TM, Guioli S, Guo W, Lalli E, Moser C, Walker AP, McCabe ER et al: An unusual member of the nuclear hormone receptor superfamily responsible for X-linked adrenal hypoplasia congenita. Nature 1994;372:635–641.
61 Wajnrajch MP, Gertner JM, Harbison MD, Chua SC, Leibel RL: Nonsense mutation in the human growth hormone-releasing hormone receptor causes growth failure analogous to the little (lit) mouse. Nat Genet 1996;12:88–90.
62 Kang S, Graham JM Jr, Olney AH, Biesecker LG: GLI3 frameshift mutations cause autosomal dominant Pallister-Hall syndrome. Nat Genet 1997;15:266–268.
63 Landis CA, Harsh G, Lyons J, Davis RL, McCormick F, Bourne HR: Clinical characteristics of acromegalic patients whose pituitary tumors contain mutant Gs protein. J Clin Endocrinol Metab 1990;71:1416–1420.
64 Layman LC, Cohen DP, Jin M, Xie J, Li Z, Reindollar RH, Bolbolan S, Bick DP, Sherins RR, Duck LW, Musgrove LC, Sellers JC, Neill JD: Mutations in gonadotropin-releasing hormone receptor gene cause hypogonadotropic hypogonadism. Nat Genet 1998;18:14–15.
65 Thomas PQ, Dattani MT, Brickman JM, McNay D, Warne G, Zacharin M, Cameron F, Hurst J, Woods K, Dunger D, Stanhope R, Forrest S, Robinson IC, Beddington RS: Heterozygous HESX1 mutations associated with isolated congenital pituitary hypoplasia and septo-optic dysplasia. Hum Mol Genet 2001;10:39–45.

66 Netchine I, Sobrier ML, Krude H, Schnabel D, Maghnie M, Marcos E, Duriez B, Cacheux V, Moers A, Goossens M, Gruters A, Amselem S: Mutations in LHX3 result in a new syndrome revealed by combined pituitary hormone deficiency. Nat Genet 2000;25:182–186.
67 Wu W, Cogan JD, Pfaffle RW, Dasen JS, Frisch H, O'Connell SM, Flynn SE, Brown MR, Mullis PE, Parks JS, Phillips JA, Rosenfeld MG: Mutations in PROP1 cause familial combined pituitary hormone deficiency. Nat Genet 1998;18:147–149.
68 Osorio MG, Kopp P, Marui S, Latronico AC, Mendonca BB, Arnhold IJ: Combined pituitary hormone deficiency caused by a novel mutation of a highly conserved residue (F88S) in the homeodomain of PROP-1. J Clin Endocrinol Metab 2000;85:2779–2785.
69 Pfaffle R, Kim C, Otten B, Wit JM, Eiholzer U, Heimann G, Parks J: Pit-1: clinical aspects. Horm Res 1996;45(suppl 1):25–28.

Sally A. Camper, Department of Human Genetics, 4301 Medical Science Research Bldg III
University of Michigan Medical School, Ann Arbor, MI 48109-0638 (USA)
Tel. +1 734 763 0682, Fax +1 734 763 7672, E-Mail donnamm@umich.edu/scamper@umich.edu

Rappaport R, Amselem S (eds): Hypothalamic-Pituitary Development. Genetic and Clinical Aspects.
Endocr Dev. Basel, Karger, 2001, vol 4, pp 13–29

Neural Crest and Pituitary Development

Heather C. Etchevers[a], *Christine Vincent*[a], *Gérard Couly*[b]

[a] Institut d'Embryologie Cellulaire et Moléculaire du CNRS et du Collège de France, Nogent-sur-Marne et
[b] Service de Stomatologie, Hôpital Necker – Enfants Malades, Paris, France

Introduction

The pituitary gland is composed of three functionally and histologically distinct regions, the anterior, intermediate and posterior lobes. These arise from two distinct embryonic primordia. The anterior (adeno)hypophysis becomes visible in humans at the end of the first month of development. It develops from an invagination of the ectoderm of the stomodeal roof, called Rathke's pouch. The intermediate lobe arises from the dorsal part of Rathke's pouch closest to the diencephalon, as has been shown in the fetal rabbit by Schimchowitsch et al. [1]. In contrast, the posterior (neuro)hypophysis pinches off of the ventral diencephalon, remaining attached by a bridge which will develop into the hypophyseal stalk.

Towards the end of the second month, the developing pituitary establishes its links to the hypothalamus by the progressive elaboration of the hypophyseal stalk, the median eminence and the venous portal circulations. The architecture of the human hypothalamo-hypophyseal system is schematized in figure 1. At the end of embryonic development, the diverse elements of the pituitary gland are grouped together within the sella turcica of the sphenoid bone complex. This chapter discusses the origins of the anlagen of the pituitary gland and their relationship to the embryological cell population, the neural crest.

Fish, amphibian, avian and mammalian animal model systems have all made contributions to understanding the multiple origins of pituitary components and inductive influences acting during embryogenesis. Our work has employed the technique of creating quail-chick chimeras, developed by Le Doua-

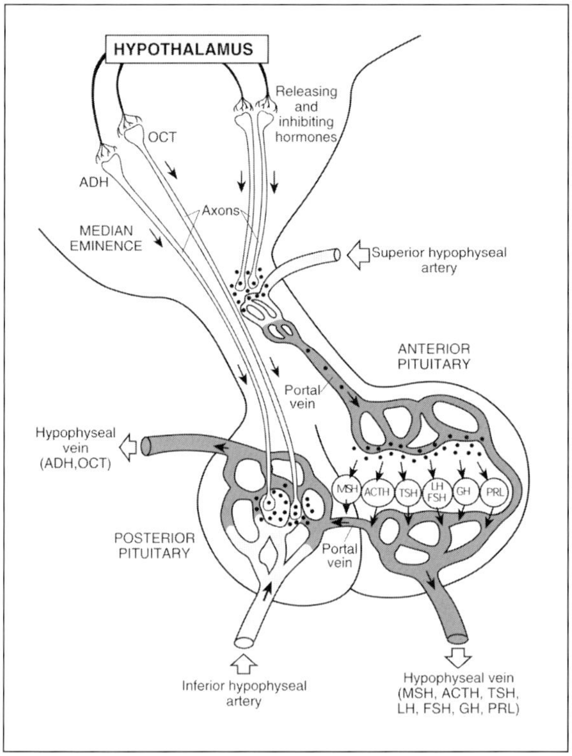

Fig. 1. Schematic diagram of the mature human hypothalamo-hypophyseal complex and its vasculature [adapted from 39]. MSH = Melanocyte-stimulating hormone; ACTH = adrenocorticotropic hormone; TSH = thyroid-stimulating hormone; LH = luteinizing hormone; FSH = follicle-stimulating hormone; GH = growth hormone; PRL = prolactin.

rin [2; reviewed in 3]. Cells from the donor species can be distinguished from those of the host by a choice of histological or immunocytochemical techniques. As there is perfect integration of the donor cells within the host, it is possible to follow their differentiation in any given location and from various time points in embryogenesis. This technique has been particularly useful in the study of the embryonic cell population known as the neural crest. Neural crest cells (NCC) delaminate from the boundary between ectoderm and the neuroepithelial plate as the latter forms a tube which will create the axis of the central nervous system. They then disperse throughout the body and participate in the development of a plethora of tissues and organs (table 1).

Using such quail-chick chimeras, we constructed a map of the presumptive territories corresponding to the adenohypophysis and neurohypophysis in the

Table 1. Derivatives of NCC in the vertebrate body

Mesectoderm	Peripheral nervous system	Endocrine cells	Pigment cells
Connective tissue of glands	*Sensory neurons of:*	*Adrenal medulla*	*All, excepting retinal pigment cells*
Pituitary	Spinal ganglia		
Lacrimal	Trigeminal (nerve V) ganglion in par	*Carotid body:*	
Salivary		Type I satellite-type cells	
Thyroid	Facial (nerve VII) ganglion	Type II satellite-type cells	
Parathyroid			
Thymus	Glossopharyngeal (nerve IX) superior ganglion	*Calcitonin-producing 'C' cells*	
Head:			
Skull (details in table 3)	Vagal (nerve X) jugular ganglion		
Odontoblasts and tooth papillae	Rohon-Beard cells (amphibians)		
Dermis			
Smooth muscle and pericytes of vascular walls	*Satellite/glial cells of:*		
	All sensory ganglia		
Connective tissue of ocular and masticatory muscles	Includes geniculate (nerve VII), otic (nerve VIII), petrosal (nerve IX) and nodose (nerve X) ganglia		
Adipose tissue in calvarium, face and ventral neck			
	All autonomic ganglia and plexi		
Heart:	Schwann cells of all peripheral nerves		
Conotruncus			
Sigmoid valves			
Smooth muscle walls of ventral aorta, aortic arches	*Autonomic neurons of:*		
	Sympathetic ganglia and plexi		
	Parasympathetic ganglia and plexi		
Trunk:			
Dorsal fins of lower vertebrates	Enteric plexi		

open neural plate [4, 5]. We also observed that cephalic NCC penetrate into the tissue of the pituitary gland. There they give rise to interstitial cells and vascular pericytes [4, 6]. Surrounding the gland, NCC also participate in the smooth muscle walls of the hypophyseal arteries and veins, and compose the bones of the anterior part of the sella turcica [7].

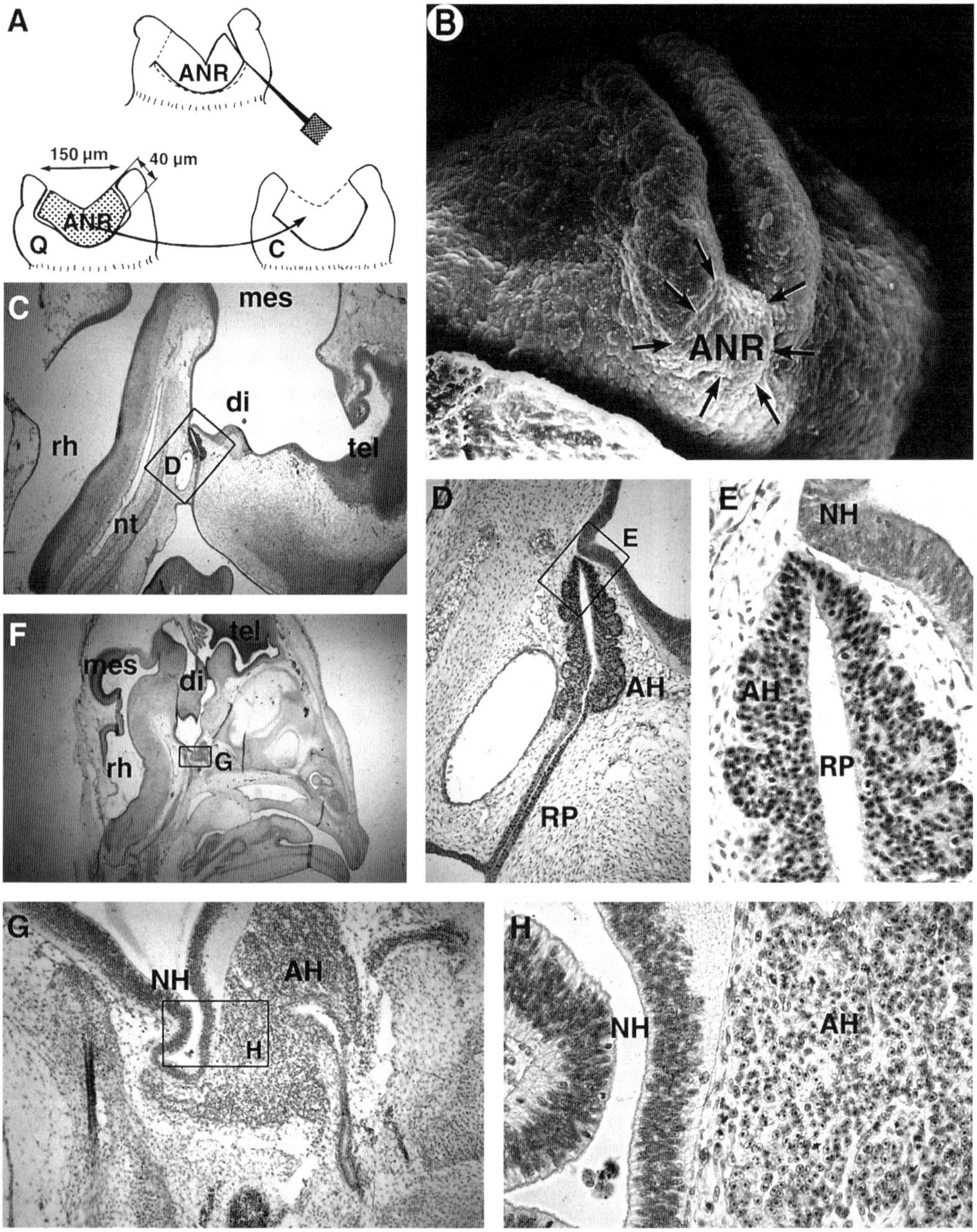

Fig. 2. Graft of quail anterior neural ridge into chicken host at 3ss leads to an adenohypophysis of donor origin [4]. *A* At the 3 somite stage, the anterior neural ridge (ANR) of a quail (Q) embryo is excised and substituted for the same region in a chicken (C) embryo

The Presumptive Adenohypophysis Is Located in the Anterior Transverse Neural Fold

When determining the location of the presumptive anterior pituitary in the avian neurula, we examined the derivatives of the anterior neural plate and its surrounding neural folds using small grafts. The sum of this data is known as a fate map, and the chimeras were constructed at a stage when the embryos had 3 somite pairs (3ss). Pieces of quail embryos on the order of 0.006 mm^2 were grafted into isochronic chicken embryos, after removing the equivalent regions of the host. Chimeric embryos were sacrificed on embryonic days (E)3, E5, E7 and E12, and the grafted cells traced.

At E3, after grafting a median piece of the anterior transverse neural fold (fig. 2A, B) one can already see that Rathke's pouch, the precursor to the adenohypophysis, is of donor origin. At E5 (fig. 2C–E) and E12 (fig. 2F–H), using the same types of grafts, the glandular cords may be seen to come from the quail graft, while the neurohypophysis and adenohypophyseal interstitial mesenchyme is contributed by the chicken host. Using similar methods, we were also able to observe that the territory corresponding to the future hypothalamus is located immediately posterior to that of the adenohypophysis. Located within the neural plate, it contains the presumptive neurohypophysis in its posterior portion [5].

In figure 3, parts of the avian fate map [8] are projected onto a neural plate seen in perspective, in order to demonstrate the morphological changes that take place over a few hours' time at the neurula stage. Similar fate maps have been constructed for zebrafish, the African clawed toad *Xenopus*, and mouse neurulas, leading to the conclusion that the underlying regional organization of the anterior neural plate is the same across all vertebrate species, including humans [reviewed in 9].

host. ***B*** The graft is seen 4 h later by scanning electron microscopy, and is delimited by arrows. ***C*** Similarly operated embryo on the 5th day of incubation, in sagittal section. Area enlarged in ***D*** indicated. Tel = Telencephalon; Di = diencephalon; Mes = mesencephalon; Rh = rhombencephalon; NT = notochord. ***D*** Rathke's pouch (RP) and the developing adenohypophysis (AH) contain quail cells, as revealed by immunohistochemistry with the quail-specific monoclonal antibody, QCPN [40]. Area enlarged in ***E*** indicated. ***E*** Close apposition of the AH with the future neurohypophysis (NH) of host origin. ***F*** Chimera on the 12th day of incubation in sagittal section. Area enlarged in ***G*** indicated. ***G*** Both AH and NH regions have developed. Area enlarged in ***H*** indicated. ***H*** Localization of quail cells by the Feulgen-Rossenbeck technique [2] shows the condensed heterochromatin of the graft present in the mature AH but not within the immediately adjacent NH.

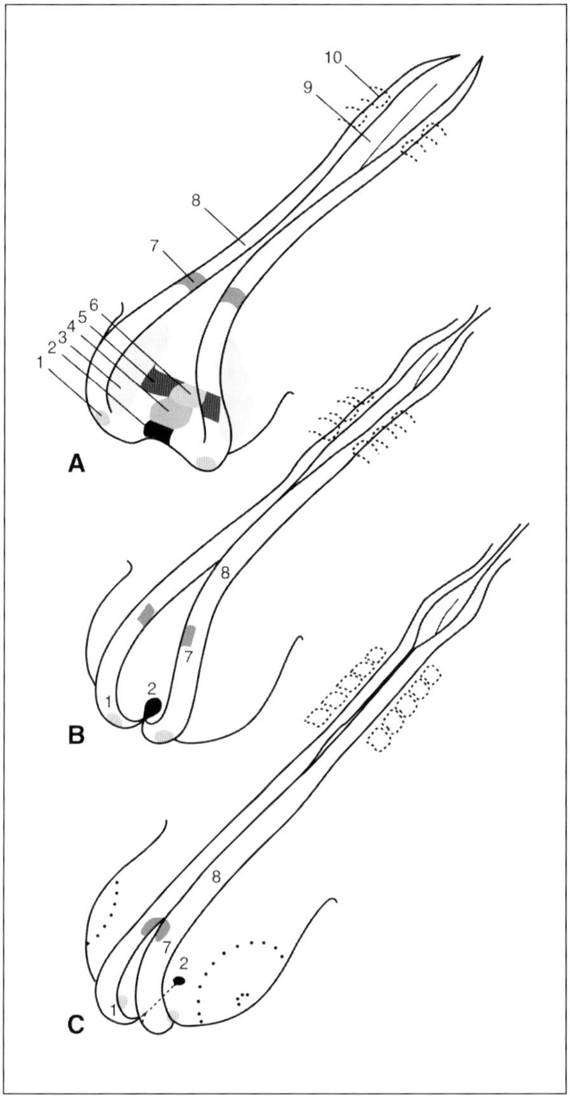

Fig. 3. Depiction of the avian neurula fate map [adapted from 8], at three different stages. ***A*** At the 2 somite stage (ss), the neural plate is relatively flat, though the neural folds are beginning to rise and approach each other dorsally. The shaded territories correspond by number to the following structures: 1 = olfactory placodes (future nasal epithelium); 2 = anterior neural ridge (future adenohypophysis); 3 = telencephalon; 4 = diencephalon (future hypothalamus); 5 = retinae; 6 = diencephalon (future neurohypophysis); 7 = epiphysis or pineal gland; 8 = neural crest; 9 = notochord; 10 = somite. ***B*** At 3ss, the fate map of the olfactory placodes, the adenohypophysis, the epiphyseal placodes and the neural crest has undergone morphological transformation as the neural folds contact each other and the new-

The Presumptive Adenohypophysis Is Subject to the Topological Deformations of the Anterior Neural Plate

Between 3ss and 10ss, the vertebrate neural plate undergoes some major morphological transformations. The anterior neural plate extends, widens, its edges rise dorsally to meet one another, and the transverse anterior edge is forced to bend ventrally (fig. 3). We examined this complex morphogenesis more closely after grafting the transverse anterior neural fold [4]. The adenohypophyseal placode territory may be seen to be transported ventral to the plane of the neuroepithelium. In the E5 chicken host, a graft of the median anterior neural fold becomes Rathke's pouch (fig. 2D). It lies underneath the diencephalon, within the posterior roof of the stomodeum. The flanking presumptive olfactory placodes roll down to lie ventral to the prosencephalic neural plate (fig. 3C). The future adenohypophysis thus comes into contact with that part of the floor of the prosencephalon from which the neurohypophysis will arise. At the same time, the hypothalamic territory, which had lain immediately adjacent inside the neural plate, slides forward to finish anterior to both of them.

Recent work in rodent models has demonstrated that separate inductive steps from the diencephalon are necessary for Rathke's pouch formation and for its differentiation [10]. The molecular signals and relays involved in these steps are beyond the scope of this chapter but are reviewed by Watkins-Chow and Camper [11], and the list continues to grow.

Colonization of the Anterior Pituitary by NCC

NCC derived from the posterior diencephalic and anterior mesencephalic neural folds begin to surround the ventral diencephalon on E3 [6]. They subsequently colonize the spaces between the glandular cords between E4 and E5 in the avian embryo (fig. 4A, B) [4, 12].

ly formed neural tube flexes forward and rolls beyond the anterior neural ridge. *C* At 5ss, the olfactory placodes now reside in an area of ectoderm which directly overlies the rhinencephalon; the adenohypophysis directly overlies the neurohypophysis, the bilaterally paired epiphyseal placodes join at the dorsal midline, and the neural folds caudal to that point will give rise to migratory NCC within a few hours. The future retinal outpocketings are depicted with dotted arcs, whereas the relative caudal regression of the adenohypophysis and joining of the immediately contiguous neural folds at the ventral midline is shown by a dashed line.

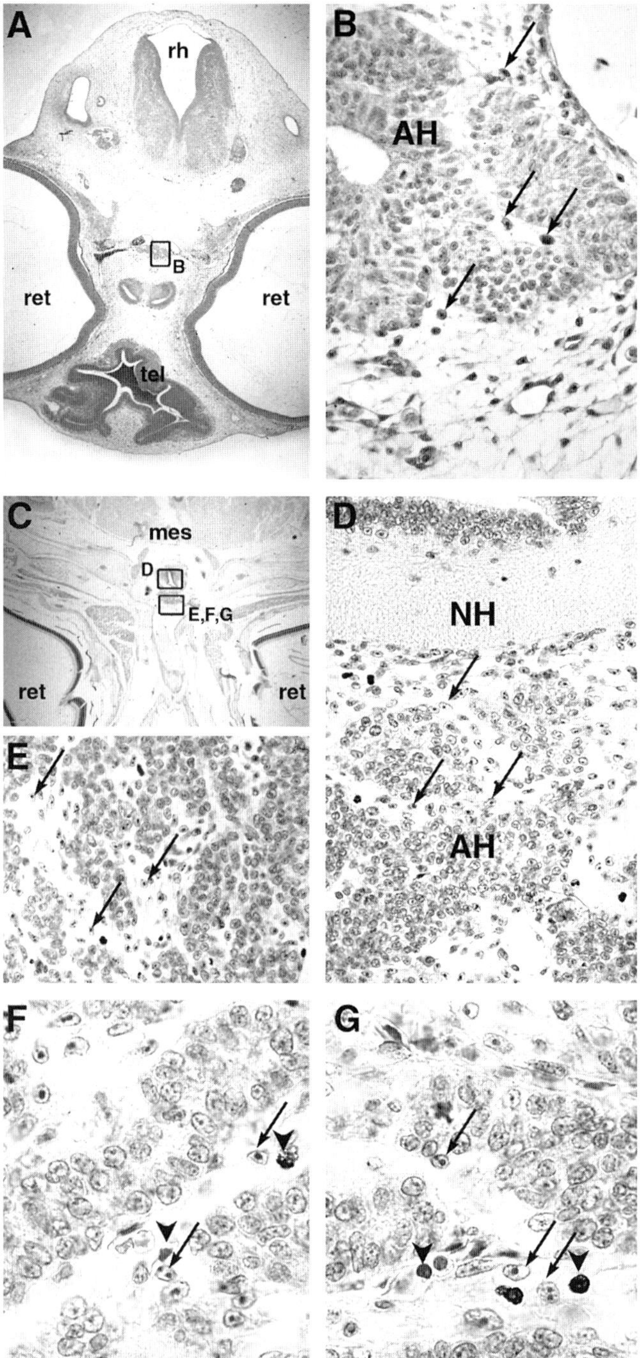
A
rh
B
ret
ret
tel
B
AH
C
mes
D
E,F,G
ret
ret
D
NH
AH
E
F
G

NCC Constitute Nonendothelial Components of the Pituitary Vascular Network

NCC which migrate into the adenohypophysis will give rise to both interstitial cells and pericytes of its capillary plexus (fig. 4C–G, 5A, B). Pericytes are smooth muscle-like cells which constitute a sheath around the capillary endothelium and regulate its local vasoconstriction or vasodilation. They express a smooth muscle-specific actin isoform, other properties reviewed by Diaz-Flores et al. [13]. Some, but not all, of the grafted NCC from the diencephalic and mesencephalic neural folds are immunoreactive for α smooth muscle actin within the adenohypophysis [14]. NCC-derived pericytes colonize the hypophyseal arteries, portal and exit veins (fig. 1). NCC are also found, less numerous, within the neurohypophysis at E12 (fig. 5B), where they appear to become exclusively pericytes since they all express α smooth muscle actin [unpubl. data].

The Adenohypophysis Does Not Develop in the Absence of the Neural Crest

We have recently shown the effects of an ablation of the neural folds of the posterior diencephalon, mesencephalon and first rhombomere, preceding the emergence of NCC [14]. Over the next day, the forebrain undergoes widespread cell death to such an extent that the retinal fields approach and cyclopia may result from their fusion (fig. 5C). Telencephalic and diencephalic tissues die from the lack of a trophic effect mediated directly or indirectly by the anteriormost NCC by way of the forming meninges.

Most of the diencephalon is eliminated after such operations (fig. 5D). Since all of the diencephalic meninges, the choroid of the eyes, and the capillary plexus of the pituitary contain NCC of anterior origin, these tissues should be affected by the absence of NCC. However, the posterior and ventral diencephalon, as well as the retinae and anterior/posterior pituitary, persist after neural fold ablations as

Fig. 4. Neural crest cells (NCC) which migrate into the adenohypophysis will give rise to both interstitial cells and pericytes of its capillary plexus. ***A*** Chimera on E7, after an isotopic graft of quail anterior mesencephalic NCC to chicken, in transverse section. Region enlarged in ***B*** indicated; ret = retina; tel = telencephalon; rh = rhombencephalon. ***B*** QCPN immunohistochemistry reveals labeled quail interstitial cells (arrows) surrounding and penetrating the adenohypophysis (AH). ***C*** Chimera on E9, after a similar graft, in transverse section and treated with the Feulgen-Rossenbeck technique. Regions enlarged in ***D***–G indicated; mes = mesencephalon. ***D***, ***E*** The interstitial tissue of the AH glandular cords is composed of the distinctive quail nuclei (arrows), indicating that it comes from the graft of mesencephalic NCC. ***F***, ***G*** Some quail NCC-derived pericytes (arrows) are located in close association with capillaries containing erythrocytes (arrowheads).

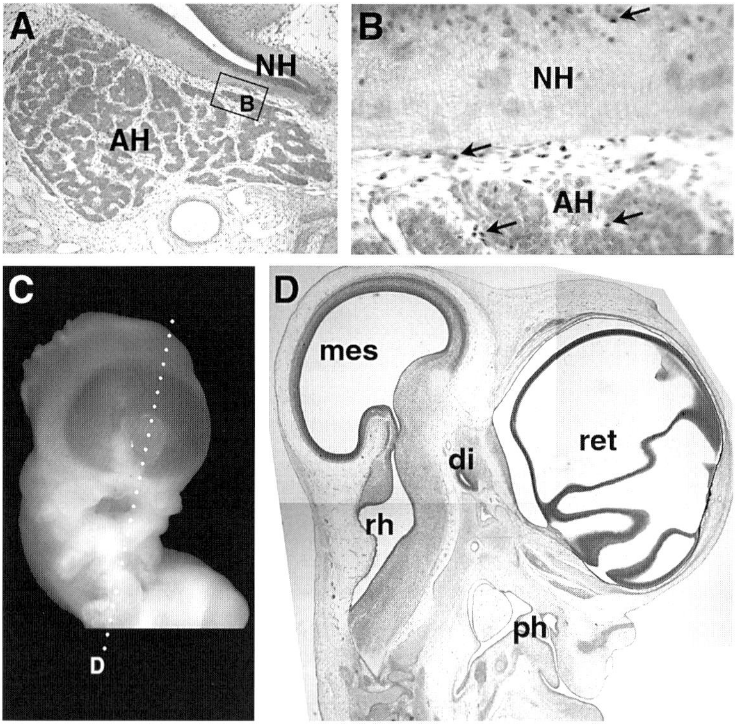

Fig. 5. NCC-derived mesenchyme which participates in pituitary gland construction is also necessary for the survival of the entire forebrain and associated structures. **A** Chick-quail chimera on 12th day of incubation, in parasagittal section with QCPN immunohistochemistry. Rectangle indicates area enlarged in **B**. **B** Grafted posterior diencephalic NCC are found to be numerous among the glandular cords and surrounding the adenohypophysis, but are fewer within the neurohypophysis, where they only contribute to the pericytic lineage. **C** Embryo in which posterior diencephalic, mesencephalic and first rhombomeric neural folds have been ablated at 5ss, preceding NCC migration. The two retinal fields have fused, yielding this E8 cyclops. Plane of section in **D** is depicted with dotted line. **D** Parasagittal section shows that the telencephalon and most of the diencephalon are absent, while only a small vestige of the posterior diencephalon and the retinae persist. The midbrain and the caudal central nervous system are unaffected by the operation. AH = adenohypophysis; di = diencephalon; mes = mesencephalon; NH = neurohypophysis; ph = pharynx; ret = retinae; rh = rhombencephalon.

described above. Isotopic grafts of the neural folds of the first rhombomere from quail, after the same long ablations in chick hosts, demonstrate that NCC of graft origin migrate rostrally. In ectopic locations, they proliferate to surround the remaining vestiges of the diencephalon, the retinae and infiltrate the adenohypophysis [14].

Nonetheless, in the most severely affected cyclops, the entire pituitary is absent, although its disappearance occurs subsequent to Rathke's pouch invagination. It takes an additional ablation of the neural folds down to the sixth rhombomere to sufficiently deprive the mesenchyme of NCC for the retinae as well as the ventral diencephalon to undergo cell death by E4 [unpubl. data]. These results suggest that the mesenchyme supplies a trophic factor, not only to the telencephalon, but also to the diencephalon and its associated outpocketings of retinae, pineal and pituitary glands, during a critical period of development which precedes vascularization.

In vitro, combinations of Rathke's pouch or head ectoderm not usually involved in pituitary development, with the ventral diencephalon, lead to induction of anterior pituitary-specific cell types in the former [15, 16]. The concomitant presence of NCC-containing mesenchyme in these avian, mammalian or mixed culture systems is necessary for Rathke's pouch survival and subsequent differentiation, in agreement with our in vivo observations.

Cephalic NCC Deficiencies May Underlie Both Congenital Hypopituitarism and Associated Human Prosencephalic and Nasofrontal Malformations

At birth, congenital hypopituitarism is evoked by symptoms such as hypoglycemia, micropenis and cryptorchidism in boys, diabetes insipidus, and/or a decrease in postnatal growth rate. A later evaluation of somatotropic, corticotropic, thyrotropic and prolactin secretions as well as neuroradiology confirm the diagnosis.

Throughout the medical literature, human pituitary deficiencies are frequently reported in association with numerous other defects of the forebrain and with premaxillary nasofrontal malformations. The associations are summarized in a nonexhaustive manner in table 2. These engendered a hypothesis that facial dysmorphologies are predictive of cephalic dysfunction [17]. The experimental data we describe above support that classic observation by showing that rostral cephalic NCC are primordial for nasofrontal and maxillary skull formation as well as prosencephalic, ocular and adenohypophyseal development in animal models. The defects and syndromes presented in table 2 are hence either themselves or are associated with cephalic neurocristopathies.

Table 2. Craniofacial and forebrain defects associated with human pituitary dysfunction

Prosencephalic malformations	Ref.	Nasofrontal, premaxillary, facial malformations	Ref.
Holoprosencephaly	17–19	Hypoplasia with hypotelorism and median cleft lip	17
Anosmia (Kallmann syndrome)	20	Iris-dental dysplasia (Rieger's syndrome	21
Optic hypoplasia; Leber disease	22, 23	Strabism	23
Septo-optic dysplasia	24, 25	Silver-Russell syndrome	26
Agenesis of the corpus callosum	22, 23, 27	Agenesis or solitary upper central and lateral incisors	19, 27, 28
		Facial clefts: lip, maxillary, palatal – isolated or in syndromes such as EEC or CHARGE association	27, 29, 30
		Maxillonasal hypoplasia (Binder syndrome)	18, 27
		Nasal alar hypoplasia (Johanson-Blizzard syndrome)	31
		Oculoauriculovertebral dysplasia (Goldenhar syndrome)	27, 32
		Congenital facial dysplasia (Möbius syndrome)	32
		Oral-facial-digital syndrome type VI (Varadi syndrome)	33, 34

The Rostral Half of the Sella turcica Derives from Mesencephalic NCC

The sphenoid bone complex consists of numerous bones. These can be segregated into two parts based on their embryological origin, demonstrated using quail-chick chimeras. The posterior region, the basipostsphenoid, is derived from the cephalic mesoderm [35]. The anterior portion, the basipresphenoid, derives from the neural crest of the mesencephalic neural folds [7]. Both portions contain cartilaginous bones. The sphenoid bone complex represents an interface between the mesodermally-derived 'chordal' skull, located in proximity to the notochord, and the 'prechordal' skull, which grows in front of the anterior end of the notochord and differentiates from NCC (table 3).

A theory presented by Gans and Northcutt [36] proposed that a 'new head' was added over the course of evolution to the 'old head' of jawless ancestors to the vertebrates. Structural adaptations afforded by NCC conferred a selective advantage on animals which were able to assume a more aggressive feeding lifestyle. The enormous contribution of the 'new head' to the vertebrate skull may be seen

Table 3. Origins of the bones of the vertebrate skull (membrane bones in italics)

Prechordal skull		Chordal skull	
Neural crest-derived		Cephalic mesoderm-derived	Somitic mesoderm-derived
Viscerocranium (face)	**Neurocranium (brain case)**		
Angular	Columella	Supraoccipital	Occipital
Dentary	Ethmoid	Sphenoid (basipost-, orbito-)	Otic capsule (pars ampullaris)
Epibranchial	*Frontal*	Otic capsule (pars canalicularis, pars cochlearis in part)	
Hyoid	*Interorbital septum*		
Jugal, quadratojugal	Otic capsule (pars cochlearis in part, parotic process)		
Meckel's cartilage	*Parietal*		
Maxilla	Sphenoid (basipre-)		
Nasal	*Squamosal*		
(Odontoblasts, tooth papillae)			
Opercular			
Palatine			
Pterygoid			
Scleral ossicles			
Superangular			
Vomer			

in the derivative bones of the NCC (fig. 6). They constitute the major part of the vertebrate head, in particular the jaws and face. We recently proposed that the telencephalon also expanded over the course of evolution due to NCC plasticity in head morphogenesis [14].

The pituitary gland lies within the boundary between the NCC-dependent and NCC-independent head, within the composite sella turcica. On the one hand, predating the development of the 'new head', pituitary development genes and some of its endocrine functions may be found in jawless chordates such as *Amphioxus* [37, 38]. On the other hand, dependence on mesenchyme for survival and differentiation of the pituitary gland argues that the organization and range of hypophyseal activity was extended in concert with the development of the cephalic NCC population in vertebrates. Like the bone by which it is encased, the pituitary gland has multiple embryonic origins, doubtless reflecting the continuous evolution of its many physiological roles over time.

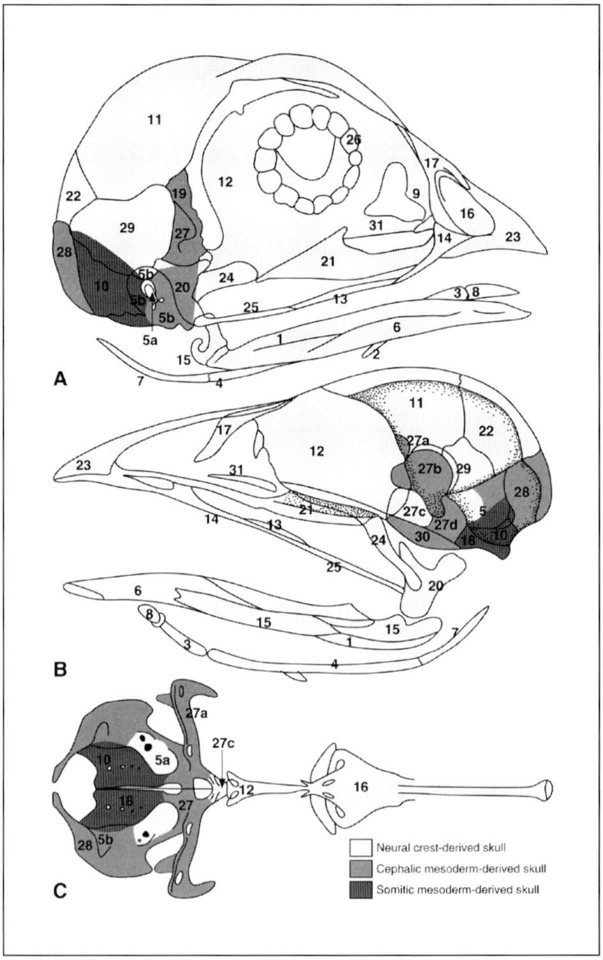

Fig. 6. Drawing of the multiple origins of the bones in the avian skull [7]. ***A*** View of the right exterior. ***B*** View of the right interior. ***C*** View from underneath. Note that the sella turcica is composed of bones of both neural crest (27c) and mesodermal (27d) origin. 1 = Angular; 2 = basibranchial; 3 = basihyal; 4 = ceratobranchial; 5 = columellar (a) and otic capsule (b); 6 = dentary; 7 = epibranchial; 8 = entoglossal; 9 = ethmoid; 10 = exoccipital; 11 = frontal; 12 = interorbital septum; 13 = jugal; 14 = maxillary; 15 = Meckel's cartilage; 16 = nasal capsule; 17 = nasal; 18 = basioccipital; 19 = postorbital; 20 = square; 21 = palatine; 22 = parietal; 23 = premaxillary; 24 = pterygoid; 25 = quadratojugal; 26 = sclerotic; 27 = sphenoid (a, orbitosphenoid; b, pleurosphenoid; c, basipresphenoid; d, basipostsphenoid); 28 = supraoccipital; 29 = squamosal; 30 = temporal; 31 = vomer.

Conclusions

The pituitary gland is made of two functional areas which occupy distinct regions within the rostral neural plate of the early vertebrate embryo. The adenohypophyseal placode is located in the medial portion of the anterior neural fold. The presumptive neurohypophysis lies within the neural plate, immediately caudal to an intervening region of the ventral hypothalamus which neighbors the neural fold. These distinct areas come together due to the topological deformations of the anterior neural plate and the surrounding head. During early development, the pituitary is ensconced in a matrix of cephalic mesoderm and NCC. These mesenchymal cells associate to create both arterial and venous aspects of the specialized vascular system of the pituitary, and in their absence, glandular survival and differentiation do not occur. Simultaneously, the sella turcica develops, of which the anterior and dorsal bones are derived from NCC while the posterior, ventral bones come from cephalic mesoderm. The neural crest population thus intervenes in an indispensable manner at multiple stages of pituitary development.

Acknowledgements

The authors warmly thank Professor Nicole Le Douarin for her long-term collaboration over the course of the work described herein as well as for her constructive criticism. They also acknowledge the technical assistance of Louis Addade, Pierre Coltey and Monique Le Thierry, and the graphical expertise of Michel Fromaget and Sophie Gournet.

References

1 Schimchowitsch S, Plant M, Klein MJ, Stoeckel ME: Ontogeny of the rabbit pituitary intermediate lobe. An ultrastructural and immunocytochemical study. Anat Embryol (Berl) 1993;187:87–97.

2 Le Douarin N: Particularités du noyau interphasique chez la caille japonaise *(Coturnix coturnix japonica)*. Utilisation de ces particularités comme 'marquage biologique' dans les recherches sur les interactions tissulaires et les migrations cellulaires au cours de l'ontogenèse. Bull Biol Fr Belg 1969; 103:435–452.

3 Le Douarin NM, Kalcheim C: The Neural Crest, ed 2. Cambridge, Cambridge University Press, 1999.

4 Couly GF, Le Douarin NM: Mapping of the early neural primordium in quail-chick chimeras. I. Developmental relationships between placodes, facial ectoderm, and prosencephalon. Dev Biol 1985;110:422–439.

5 Couly GF, Le Douarin NM: Mapping of the early neural primordium in quail-chick chimeras. II. The prosencephalic neural plate and neural folds: Implications for the genesis of cephalic human congenital abnormalities. Dev Biol 1987;120:198–214.

6 Etchevers HC, Vincent C, Le Douarin NM, Couly GF: The cephalic neural crest provides pericytes and smooth muscle cells to all blood vessels of the face and forebrain. Development 2001;128: 1059–1068.

7 Couly GF, Coltey PM, Le Douarin NM: The triple origin of skull in higher vertebrates: A study in quail-chick chimeras. Development 1993;117:409–429.
8 Couly G, Le Douarin NM: The fate map of the cephalic neural primordium at the presomitic to the 3-somite stage in the avian embryo. Development 1988;103(suppl):101–113.
9 Rubenstein JL, Shimamura K, Martinez S, Puelles L: Regionalization of the prosencephalic neural plate. Annu Rev Neurosci 1998;21:445–477.
10 Takuma N et al: Formation of Rathke's pouch requires dual induction from the diencephalon. Development 1998;125:4835–4840.
11 Watkins-Chow DE, Camper SA: How many homeobox genes does it take to make a pituitary gland? Trends Genet 1998;14:284–290.
12 Le Lièvre CS, Le Douarin NM: Mesenchymal derivatives of the neural crest: Analysis of chimaeric quail and chick embryos. J Embryol Exp Morphol 1975;34:125–154.
13 Diaz-Flores L, Gutierrez R, Varela H, Rancel N, Valladares F: Microvascular pericytes: A review of their morphological and functional characteristics. Histol Histopathol 1991;6:269–286.
14 Etchevers HC, Couly G, Vincent C, Le Douarin NM: Anterior neural crest is required for forebrain viability. Development 1999;126:3533–3543.
15 Treier M, Gleiberman AS, O'Connell SM, Szeto DP, McMahon JA, McMahon AP, Rosenfeld MG: Multistep signaling requirements for pituitary organogenesis in vivo. Genes Dev 1998;12:1691–1704.
16 Gleiberman AS, Fedtsova NG, Rosenfeld MG: Tissue interactions in the induction of anterior pituitary: Role of the ventral diencephalon, mesenchyme, and notochord. Dev Biol 1999;213:340–353.
17 De Myer W, Zeman W, Palmer CG: The face predicts the brain: Diagnostic significance of median facial anomalies for holoprosencephaly (arhinencephaly). Pediatrics 1964;34:256.
18 Binder KH: Dystostosis maxillo-nasalis, ein arhinencephaler Missbildungskomplex. Dtsch Zahnärztl Z 1962;17:438.
19 Fleming P, Nelson J, Gorlin RJ: Single maxillary central incisor in association with mid-line anomalies. Br Dent J 1990;168:476–479.
20 Bardin CW, Ross GT, Rifkind AB, Cargille CM, Lipsett MB: Studies of the pituitary Leydig cell axis in young men with hypogonadotropic hypogonadism and hyposmia: Comparison with normal men, prepubertal boys, and hypo-pituitary patients. J Clin Invest 1969;48:2046–2056.
21 Sadeghi-Nejad A, Senior B: Autosomal dominant transmission of isolated growth hormone deficiency in iris-dental dysplasia (Rieger's syndrome). J Pediatr 1974;85:644.
22 Gendrel D, Chaussin JL, Job JC: Les hypopituitarismes congénitaux par anomalie de la ligne médiane. Arch Fr Pediatr 1981;38:227–232.
23 Liapi C, Ginisty D, Chaussain JL, Job JC: Les malformations dentaire et faciales associées à l'insuffisance hypophysaire en hormone de croissance. Arch Fr Pediatr 1985;42:829–833.
24 De Morsier G: Etudes sur les dysraphies craniocéphaliques. III. Agénésie du septum lucidum avec malformation du tractus optique: La dysplasie septo-optique. Schweiz Arch Neurol Neurochir Psychiatr 1956;77:267–292.
25 Hoyt WT, Kaplan SL, Grumbach MM: Septo-optic dysplasia with hypopituitary dwarfism. Lancet 1970;i:893.
26 Kotilainen J, Holtta P, Mikkonen T, Arte S, Sipila I, Pirinen S: Craniofacial and dental characteristics of Silver-Russell syndrome. Am J Med Genet 1995;56:229–236.
27 Ajacques JC, David M, Disant F, Sann L, Bey-Omar F: Malformations de l'étage moyen de la face et déficit hypothalamo-hypophysaire. Pédiatrie 1982;37:417–432.
28 Rappaport EB, Ulstrom RA, Gorlin RJ, Lucky AW, Colle E, Miser J: Solitary maxillary central incisor and short stature. J Pediatr 1977;91:924–928.
29 Rudman D, Davis T, Priest JH, Patterson JH, Kutner MH, Heymsfield SB, Bethel RA: Prevalence of GH deficiency in children with cleft lip and palate. J Pediatr 1978;93:378–382.
30 Goumy P, Dalens B, Malpuech G: Association d'une dysraphie de la ligne médiane et d'une insuffisance hypophysaire congénitale avec micropenis et hypoglycémie néonatale. J Pediatr 1978;93:551–559.
31 Johanson AJ, Blizzard RM: A syndrome of congenital aplasia of the alae nasi, deafness, hypothyroidism, dwarfism, absent permanent teeth, and malabsorption. J Pediatr 1971;79:982–987.

32 Couly G, Rappaport R, Brauner R et al: Association of facial malformations with primary hypopituitarism as possible incidence for a common developmental defect of prosencephalic neural crest derivatives. Pediatr Res 1982;16:886–907.
33 Varadi V, Szabo L, Papp Z: Syndrome of polydactyly, cleft lip/palate or lingual lump, and psychomotor retardation in endogamic gypsies. J Med Genet 1980;17:119–122.
34 Munke M, McDonald DM, Cronister A, Stewart JM, Gorlin RJ, Zackai EH: Oral-facial-digital syndrome type VI (Varadi syndrome): Further clinical delineation. Am J Med Genet 1990;35:360–369.
35 Couly GF, Coltey PM, Le Douarin NM: The developmental fate of the cephalic mesoderm in quail-chick chimeras. Development 1992;114:1–15.
36 Gans C, Northcutt RG: Neural crest and the origin of vertebrates: A new head. Science 1983;220:268–274.
37 Candiani S, Pestarino M: Expression of the tissue-specific transcription factor Pit-1 in the lancelet, *Branchiostoma lanceolatum*. J Comp Neurol 1998;392:343–351.
38 Glardon S, Holland LZ, Gehring WJ, Holland ND: Isolation and developmental expression of the amphioxus Pax-6 gene (AmphiPax-6): Insights into eye and photoreceptor evolution. Development 1998;125:2701–2710.
39 Berne RM, Levy MN: Principles of Physiology, ed 2. St Louis, Mosby-Year Book, 1996.
40 Lance-Jones CC, Lagenaur CF: A new marker for identifying quail cells in embryonic avian chimeras: A quail-specific antiserum. J Histochem Cytochem 1987;35:771–780.

Dr. Heather C. Etchevers, Institut d'Embryologie Cellulaire et Moléculaire du CNRS et du Collège de France, 49bis avenue de la Belle Gabrielle, F–94736 Nogent-sur-Marne Cedex (France)
Tel. +33 1 45 14 15 15, Fax +33 1 48 73 43 77, E-Mail etchever@infobiogen.fr

Rappaport R, Amselem S (eds): Hypothalamic-Pituitary Development. Genetic and Clinical Aspects.
Endocr Dev. Basel, Karger, 2001, vol 4, pp 30–47

Perspectives on the Molecular Basis of Developmental Defects in the Human Pituitary Region

Serge Amselem

Institut National de la Santé et de la Recherche Médicale (Unité 468), Hôpital Henri-Mondor, Créteil, France

Introduction

The mature pituitary is an endocrine gland which consists of the neurohypophysis (or posterior pituitary) and the adenohypophysis. The anterior pituitary, part of the adenohypophysis, is composed of five differentiated cell types – somatotropes, lactotropes, thyreotropes, gonadotropes and corticotropes – producing six hormones, i.e. growth hormone (GH), prolactin, thyroid-stimulating hormone (TSH), luteinizing hormone (LH) and follicle-stimulating hormone (FSH) and adrenocorticotropin (ACTH). Elegant investigations into the development of the anterior pituitary performed in various species have led to the conclusion that this gland derives from the most anterior part of the neural plate [1–4]. During the developmental process, pituitary cell-type specification has been shown to result from several signal-dependent transcriptional cascades involving a large number of genes (fig. 1), the temporal and spatial expression patterns of which are so accurately regulated [reviewed in 6–11] that any alteration of these programs might have deleterious consequences on the proper development of the pituitary gland.

In humans, developmental defects of the pituitary region are rare disorders characterized by an extreme variability in expression of the disease phenotype. Patients usually present with clinical features revealing a deficit in one or several anterior pituitary hormones, the latter condition being referred to as combined

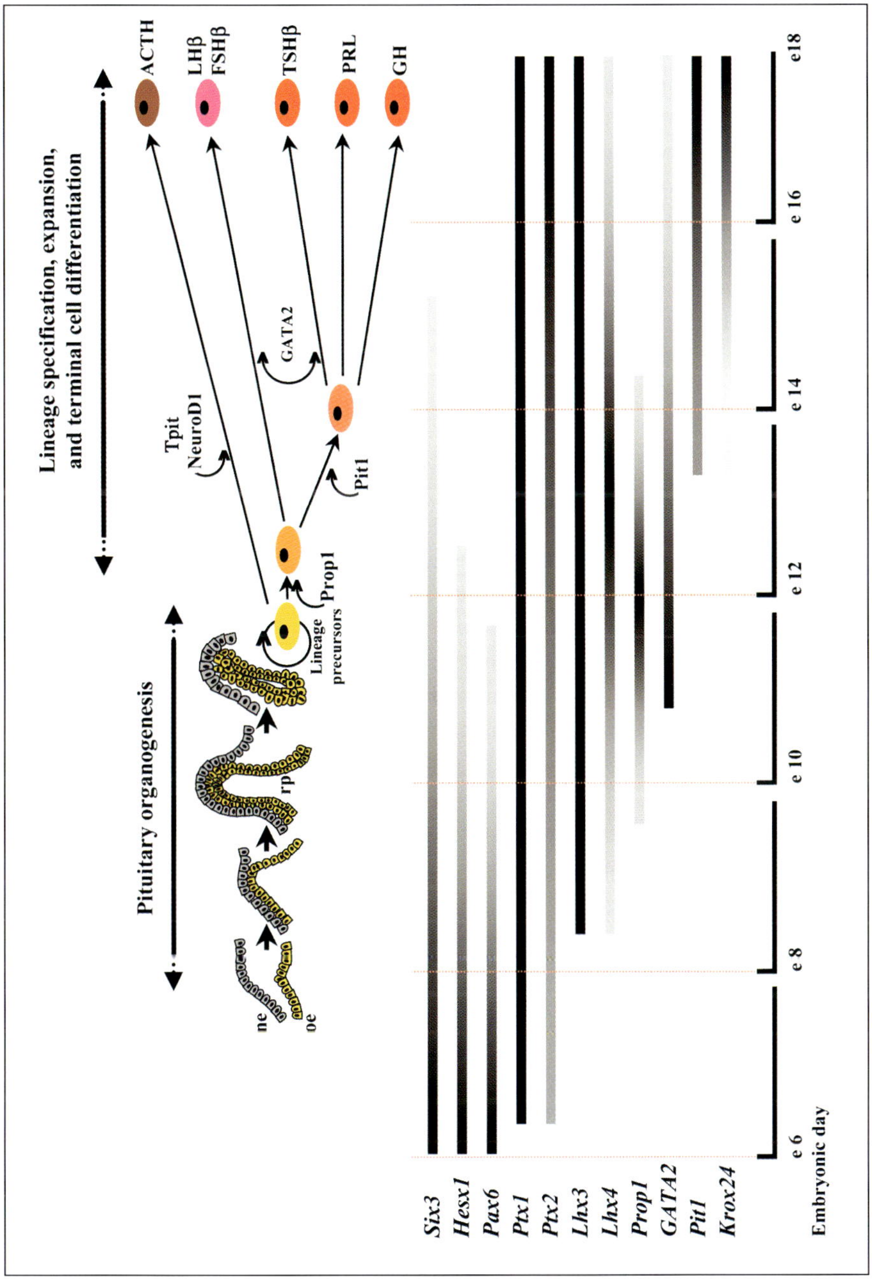

Fig. 1. Models for organogenesis of the anterior pituitary gland in the mouse and cell lineage determination by a cascade of transcription factors. oe = oral ectoderm; ne = neural epithelium; rp = Rathke's pouch [adapted from 5–8].

pituitary hormone deficiency (CPHD); the patients may also carry other developmental defects involving various extrapituitary structures, thereby defining syndromic CPHDs. In several cases, specific biological tests can also identify hormonal deficits that are clinically asymptomatic. On MRI examination, the morphology of the hypothalamic-pituitary region may considerably vary among patients, some displaying a normal morphology, others presenting with different kinds of abnormalities including those of the anterior pituitary which can be aplastic, hypoplastic or enlarged. Such phenotypic features clearly support the existence of a genetic heterogeneity associated with this condition. This hypothesis has been confirmed by the description of various modes of inheritance of the disease phenotype in the rare familial CPHD cases (i.e. autosomal recessive, autosomal dominant or X-linked). The first molecular studies have also shown that, in these Mendelian disorders, individuals with identical disease-causing mutations may differ phenotypically, and, conversely, patients with mutations in different genes may display similar disease phenotypes (see below). Altogether, these observations partly account for the difficulties in identifying mutated genes underlying such developmental anomalies.

This review focuses on the general strategies currently used to decipher the molecular bases of Mendelian disorders such as CPHD of genetic origin. By summarizing illustrative approaches that turned out to be successful for identifying genes involved in CPHD in humans, this article considers predictable obstacles specific to the investigation of these rare and heterogeneous conditions.

General Approaches for Identifying the Molecular Basis of Developmental Defects of the Pituitary Region in Humans

Position-Independent Candidate Gene Approaches

Efforts to identify genes involved in developmental defects of the pituitary region in humans have primarily relied on the evaluation of candidate genes. In the case of CPHD, the candidates could be identified by virtue of their known or expected roles in normal and pathological pituitary development in various species, without any knowledge on the chromosomal location of the corresponding human disease-causing gene. Such a position-independent candidate gene approach, which is classically considered as relatively risky by geneticists, has proven to be especially valuable for identifying genes implicated in CPHDs.

The differentiated anterior pituitary structure is the result of the activation of sequential series of molecular events that take place during anterior pituitary embryogenesis. Several studies performed in mice led to the characterization of numerous proteins involved in the differentiation of the five pituitary cell types

[6–11]. Most of these proteins, which share functional and structural characteristics, belong to the large family of homeodomain transcription factors (fig. 1), thereby suggesting their functional importance in normal pituitary development.

Mouse Models: Naturally Occurring Mutants and Gene Targeting

Mouse models can be used in different ways for dissecting physiological and pathological developmental pathways in humans. In the field of CPHD, such models have yielded direct clues for identifying several human genes that play a key role in the occurrence of developmental pituitary defects [10].

Naturally Occurring Mutants

The study of mouse models carrying naturally occurring mutations demonstrated the allelic and locus heterogeneity associated with CPHD: the dwarf phenotype of two strains, named *Snell* and *Jackson*, with a disease phenotype transmitted as an autosomal recessive trait and characterized by an absence of GH, prolactin and TSH gene expression, associated with pituitary hypoplasia, has been shown to result from different mutations in the *Pit1* gene [12]. The corresponding protein, a member of the POU family of transcription factors, is involved in transcriptional activation of GH, prolactin, TSH and GH-releasing hormone receptor genes, as well as in the differentiation and proliferation of somatotrophs, lactotrophs and thyreotrophs [12, 13]. Strikingly, a similar phenotype has been identified in a third murine strain named *Ames*. This nonallelic mutation, characterized by an absence of detectable *Pit1* transcripts in developing pituitaries [14], was shown to result from a defect in the gene encoding Prophet of Pit1 (Prop1) [15], a novel pituitary-specific *paired-like* homeodomain transcription factor expressed in the primordium of the anterior pituitary, the Rathke's pouch. These observations have prompted several groups to search for mutations in the human orthologous genes in patients presenting with similar disease phenotypes. Following a position-independent candidate gene approach, several mutations in *POU1F1* (the gene orthologous to the mouse *Pit1* gene) have been reported in patients with familial or sporadic CPHD [16–18]. In the familial forms, depending on the nature of the *POU1F1* mutation, the disease phenotype is transmitted either as an autosomal recessive or dominant trait. The first human *PROP1* mutations have been identified more recently [19, 20]. The patients displayed a disease phenotype similar to that of POU1F1-deficient patients (i.e., GH, prolactin and TSH deficiency associated with a hypoplastic anterior pituitary). However, the large majority of PROP1-deficient patients are also gonadotropin-deficient, a phenotypic feature that is in keeping with the delayed gonadotrope determination documented in *Ames* mice. Similarly, another spontaneous mouse mutation, named *Little*, valuably served as a model to identify the first human mutations in the GH-releasing hormone receptor gene (*GHRHR*) [21–23]. This strain, which carries a missense mutation within the *Ghrhr* gene [24, 25], displays a recessively inherited disease phenotype characterized by postnatal growth retardation – which was found to be insensitive to exogenous GH-releasing hormone – and a hypoplastic pituitary with a depletion of somatotroph cells and low prolactin content.

Mouse Gene Targeting

Genetically engineered mouse models with pituitary failure have also provided candidate genes for the investigation of candidate disease phenotypes in humans. This situation was illustrated by the study of mice carrying a homozygous *Hesx1* null allele [26]. The *Hsex1* gene, which encodes a transcription factor containing a homeodomain, is expressed in prospective forebrain tissue during early mouse development, but later, the expression is restricted to Rathke's pouch. As expected from this expression pattern which begins earlier than *Prop1*, *Hesx1–/–*, mice exhibit a severe and complex phenotype, with 75% of the animals dying shortly after birth [26]. The phenotypic abnormalities, combining forebrain midline defects with pituitary dysplasia, are reminiscent of those reported in human septo-optic dysplasia. The subsequent investigation of a large number of patients with various pituitary disorders identified *HESX1* mutations in some of them [26, 27]. Following similar gene targeting experiments, Sheng et al. [28, 29] showed that *Lhx3* and *Lhx4*, two LIM-homeodomain transcription factors, play a crucial role in the formation and expansion of Rathke's pouch. In homozygous mutants generated by targeted disruption of the *Lhx3* gene, Rathke's pouch failed to grow and differentiate, the mice lacking both the anterior and intermediate lobes of the pituitary, except for the presence of a small number of corticotropes that failed to proliferate [28]. In mice, targeted disruption of *Lhx4* resulted in a similar pituitary phenotype with a reduction in terminally differentiated pituitary cell lineages, leading to pituitary hypoplasia [29]. The study of Lhx3/Lhx4 double mutants demonstrated the redundant control of these two factors over formation of the definitive pouch [29]. Recently, the first human *LHX3* mutations have been identified in patients presenting with a new syndrome revealed by CPHD involving all but one (ACTH) anterior pituitary hormone [30] (see below).

It is likely that the description of CPHD phenotypes in murine strains generated by targeted disruption of other transcription factors expressed in hypothalamic or pituitary cells [reviewed in 10], like Gsh1 [31] and Titf1 [32, 33] or Krox-24 [34] and Otx1 [35], will provide a basis for the molecular characterization of other types of combined pituitary hormone deficiencies in humans.

Candidate Disease Genes Suggested by Their Expression Pattern and/or Putative Function

In theory, all the genes that are specifically expressed in the embryonic territory from which pituitary originates may be considered as good candidates for developmental abnormalities of this region. It is indeed possible to accurately determine the spatial and temporal expression pattern of a given gene by means of in situ hybridization of mRNAs in mouse embryonic sections. With this approach, an increasing number of transcripts encoding homeodomain-containing factors have been identified at different stages of pituitary organogenesis [6, 36]. Other studies [reviewed in 7] have shown the expression of multiple signaling molecule gradients during pituitary development. These transient signaling gradients are, in turn, essential to the proper expression of several of the abovementioned pituitary-specific factors: for instance, the opposing expression gradients of FGF8 (dorsal → ventral) and BMP2 (ventral → dorsal) during pituitary organogenesis are essential to the generation of overlapping patterns of transcription factor expression [37, 38].

A different approach based on the systematic analysis of a cDNA library generated from total RNA extracted from pituitaries recently permitted the identification of more than 3,000 different pituitary-specific sequences including 635 novel expressed sequence tags (ESTs) [39]. Although these sequences were isolated from adult pituitaries, it seems reasonable to consider at least some of them as candidates for developmental defects of the pituitary gland, since it is established that other genes like *POU1F1* [40], *PROP1* [19, 20] or *LHX3* [30, 41], which play a crucial role in pituitary development, do remain expressed in the adult pituitary.

Finally, most of the factors that have proven to play a major role in pituitary development act through direct protein-protein interactions. These interactions are of crucial importance for the proper pituitary cell-type determination process: depending on the factors and cell type involved, synergistic or inhibitory protein-protein interactions have been identified. They include not only contacts between pituitary-specific transcription factors [reviewed in 6], but also interactions with other classes of molecules [6, 42a]. The identification of the different proteins involved in the differentiation pathway of each pituitary cell type and the analysis of their expression pattern should provide key information which, in turn, might also suggest new candidates for developmental defects of the pituitary. Just recently, this approach led to the identification of mutations in *TPIT*– a novel T box factor cooperating with PITX homeoproteins – in patients with isolated ACTH deficiency [42b].

Approaches Based on the Determination of the Chromosomal Location of the Disease Gene

If a genetic etiology is suspected, in the absence of any relevant information about the pathogenesis of the pituitary defect documented in a patient, the identification of the chromosomal region involved in the disease phenotype represents the first important step towards the isolation of the disease-causing gene(s). Such chromosomal mappings allow the subsequent cloning of genes by means of two main approaches: first, sequence examination of the identified region may reveal the presence of gene(s) considered as good candidate(s) for the disease phenotype on the basis of their known or putative function and/or their expression pattern; this so-called 'positional candidate gene approach' has already proven to be extremely efficient in the identification of numerous disease genes in humans. Second, sequence analysis may also fail to identify such candidates. In this latter case, an approach, named 'positional cloning', will aim at identifying the genes located in the critical chromosomal region, prior to their fine analysis in the patients' DNA samples compared to control samples; this approach is less and less used by geneticists since the recent major advances in knowledge of the human genome [43] (see below). Although no mutated gene responsible for a pituitary defect has so far been isolated in humans on the basis of its chromosomal location, two main approaches may reveal valuable to identify new loci involved in such developmental defects.

Chromosome Abnormalities

As mutations within genes expressed during development are expected to result in defects of various organs, the disease phenotype of patients presenting with CPHD should be carefully specified, with particular attention to examination of extrapituitary structures. Any unexplained association of symptoms or malformations (i.e. syndromic CPHDs) should motivate the search for a chromosome abnormality. In fact, translocations, deletions, inversion, duplications or other cytogenetic abnormalities directly indicate the location of the disease-causing gene(s), with a precision that depends on the size of the genomic region involved in the rearrangement. With this approach, several candidate loci have been identified in patients who presented with CPHD associated with a chromosome abnormality [44–47].

Genetic Linkage Analyses

Linkage analysis is based on the simple concept that a DNA marker located close to a disease gene is more likely to be inherited with this gene than another gene located at a more distant site on the same or another chromosome. In theory, the two following whole-genome scanning approaches, classically used to isolate genes involved in Mendelian disorders, might also be relevant for identifying the molecular bases of pituitary developmental defects.

The first approach is the genotyping of several hundreds of highly polymorphic multiple-allelic DNA markers dispersed throughout the entire genome – namely microsatellite markers – in a large series of families affected by a genetic disorder. The intrafamilial segregation of these markers, that are easily assayed by PCR, is analyzed in all affected families, in which the genotype at each polymorphic locus is determined in each family member (i.e. patients *and* healthy subjects). If, within a family, the patients share identical polymorphic alleles at a given locus, whereas unaffected relatives carry a different genotype at the same locus, the corresponding chromosomal region may contain the gene underlying the disease in this family. To confirm this assumption, it is generally necessary to perform similar investigations in additional families concerned by the same disease. This approach requires several conditions to be fulfilled: (1) The disease has to be homogeneous enough – from the genetic viewpoint – to allow the identification of the causative gene. (2) The parents of affected children must carry different polymorphisms, so that it will be possible to distinguish the alleles inherited from each parent. In practice, familial pituitary defects are rare conditions which are expected to be so heterogeneous at the genetic level that the first condition cannot easily be met. The intrafamilial variability of the disease phenotype associated with pituitary defects further underlines the difficulties in establishing a positive genetic linkage in these families. However, as shown for many other Mendelian disorders, the investigation of a single affected kindred, large enough to allow linkage studies to be performed, could lead to the identification of the underlying mutated gene in that particular family. As mentioned above, this analysis would, however, require the participation of healthy relatives.

In contrast, the second approach, which can be used to map loci involved in autosomal recessive conditions, theoretically requires only the participation of the patients. This approach, named 'homozygosity mapping', relies on the fact that consanguineous patients are expected to carry homozygous markers at the disease locus [48]. As in conventional linkage analyses, it may be preferable to perform these investigations in related patients, because of the important genetic heterogeneity associated with pituitary defects [49]. However, in contrast to conventional linkage, homozygosity mapping has proven to be a powerful way to identify a disease locus with the participation of a limited number of patients.

Table 1. General strategies for identifying genes underlying developmental defects of the pituitary region

1	Position-independent candidate gene approaches Mouse models: spontaneous mutations/genetically engineered models Candidates suggested by their expression pattern Candidates suggested by their putative function
2	Approaches based on the chromosomal location of the disease gene Chromosome abnormalities Genetic linkage analyses
3	Use of new sequence data – computational biology

New Directions

A combination of these different approaches (listed in table 1) may facilitate the identification of the mutated gene. For instance, in familial cases, several candidate genes can be first assessed by means of linkage analysis using polymorphic DNA markers located in the vicinity of these genes. An exhaustive analysis of all exons and intronic boundaries of the candidate gene will be performed only in case of positive linkage between the genomic region of interest and the disease phenotype. In addition, such linkage studies can take great advantage of the considerable conservation of synteny between human and mouse chromosomes. This strategy applies especially to the investigation of disorders like CPHDs in which similar disease phenotypes may result from mutations in different genes. Indeed, patients displaying a deficit in GH, prolactin and TSH may carry defects in *POU1F1* or in *PROP1*; this is especially the case with prepubertal children in whom peak levels of gonadotropins are not interpretable. Similarly, a deficiency in GH, prolactin, TSH, FSH and LH may result from mutations in *PROP1* or in *LHX3* [30]. Lastly, and most importantly, these approaches, the main limitations of which are mentioned below, are currently challenged by the vast amount of data being produced by the Human Genome Project and other sequencing efforts [43, 50a, 50b, 50c, 50d, 50e].

The Use of New Sequence Information

The particularly large production of new sequences over the past year now makes it possible to isolate full-length genes or cDNAs by computational methods alone (i.e. *in silico* approaches). These data should greatly facilitate the direct evaluation of candidate genes in many as yet unexplained human diseases, just by sequencing the regions of interest. This will be the case especially for pituitary defects which are particularly difficult to explore by means of more conventional approaches because either the disease is rare, the structure of the affected family precludes any kind of linkage analyses, or the patients represent sporadic cases.

Although genetic maps based on microsatellite markers have been particularly useful in identifying disease genes, a new wave of much more densely spaced markers is now becoming available. These bi-allelic markers, named single nucleotide polymorphisms (SNPs), which occur once every 1,900 base pairs in the human genome, promise to be particularly helpful in dissecting genetically complex phenotypes. Such a huge number of SNPs per genome should allow the fine mapping of disease traits, in spite of the fact that these markers have only two alleles [51, 52a, 52b].

From Multispecies Sequence Information to Mouse/Human Functional Data

These sequencing efforts, which not only apply to the human and mouse genomes but also to numerous other species, allow comparisons to be made between species; this comparative genomics should facilitate further gene identification and elucidation of gene functions [53, 54]. As almost all human genes are also present in mice, it is theoretically possible to design mouse models for virtually all human Mendelian disorders. Therefore, the identification of a gene mutation in a patient displaying a developmental defect of the hypothalamic-pituitary region may encourage the generation of engineered mouse mutants in the orthologous sequence. The study of various mouse models carrying different kinds of mutations in the context of different genetic backgrounds could then provide important insights into the pathophysiology of the human disorder. It is for instance now possible to induce a targeted disruption of a given gene in a tissue- and temporal-specific manner [55, 56]. These conditional targeting experiments represent a powerful way to study the specific role of a gene in almost any fetal or adult tissue. In addition, the use of targeted mutagenesis in the mouse may be extremely valuable to investigate the role of a gene which germinal knockout leads to developmental or perinatal lethality. The generation of such animal models, in which the expression of a targeting event is controlled both temporally and spatially, could also be valuable for identifying the physiological functions of pituitary-specific transcription factors that are known to play important roles in pituitary development and the expression of which is strikingly maintained in adult pituitary (e.g. LHX3 [30, 41] or PROP1 [19, 20]).

Table 2. Spontaneous and engineered mouse models for pituitary defects in human patients

Genes	Mouse strains			Human mutations							
	name	molecular defect	ref.	GH	Prl	TSH	FSH	LH	ACTH	associated malformations	ref.
GHRHR	Little	R60G	24, 25	–	±	+	+	+	+	No	21–23
POU1F1	Jackson	Rearrangement	12	–	–	±	+	+	+	No	16–18
	Snell	Y261C									
PROP1	Ames	S83P	15	–	–	–	–	–	±	No	19, 20
LHX3		KO	28	–	–	–	–	–	+[a]	Yes	30
HESX1		KO	26	–	–	–	–	–	–	Yes	26, 27

GH = Growth hormone; Prl = prolactin; TSH = thyroid-stimulating hormone; FSH = follicle-stimulating hormone; LH = luteinizing hormone; ACTH = adrenocorticotropin; KO = knockout; – = severe deficit; + = normal hormonal levels; ± denotes hormonal values that are either normal, subnormal, or severely decreased.

[a] The report (in 1 patient) of an enlarged anterior pituitary which occurred secondarily leaves open the possibility that, like in PROP1-deficient patients, patients with *LHX3* mutations may present secondarily with a deficit in ACTH, by means of a so far unexplained phenomenon associated with the regression of the pituitary mass.

Human/Mouse Differences and Limitations of Extrapolation from Mouse Models

Although, as detailed above and summarized in table 2, mouse strains carrying mutations that are either spontaneous or generated by gene targeting have substantially contributed to a better understanding of the developmental defects of the pituitary gland in humans, these models have certain limitations. Indeed, as shown in several other pathological conditions, mouse models may display a disease phenotype not identical to that reported in humans. This situation has been clearly illustrated by the study of *LHX3*, *PROP1* and *HESX1* in both species. Although the disease phenotype of patients with *LHX3* mutations recalls the phenotype of mice carrying a homozygous targeted disruption of *Lhx3*, it is worth noting that, in mice, the phenotype is much more severe (table 3), leading to pituitary aplasia and perinatal death of unknown origin [28]; in addition, in these mice, corticotropes fail to proliferate. In contrast, the pituitary of patients with *LHX3* mutations was found to be either hypoplastic or dramatically enlarged, and the patients presented with a CPHD phenotype involving all anterior pituitary hormones except adrenocorticotropin [30]; in addition, the 4 patients so far identified with *LHX3* mutations also presented with an as yet unexplained rigid cervical spine leading to limited head rotation. Similarly, the disease phenotype of the *Ames* dwarf strain, which carries a homozygous *Prop1* mutation, is characterized by a severe hypoplasia of the anterior pituitary, whereas the pituitary size from

Table 3. Phenotypic features associated with *LHX3* mutations in mice and human patients

Phenotypic features	Mice (Lhx3–/–) [28]	Humans [30]
Viability	Perinatal lethality	Viable
Pituitary size	Pituitary aplasia	Pituitary hypo/hyperplasia
Corticotrope function	Failure of corticotropes to proliferate	Normal
Associated malformations		Neck abnormalities

Table 4. Main mechanisms accounting for limitations of extrapolation from mouse models to human disorders

1	Nature of the disease-causing mutations
2	Species-specific embryonic expression patterns
3	Unique or redundant biochemical pathways
4	Role of genetic backgrounds
5	Stochastic events during embryonic development
6	Modification of the disease phenotype with time

patients with *PROP1* mutations is extremely variable, and may be reduced, normal or enlarged, even in patients carrying the same *PROP1* defect [57]. In the same way, targeted disruption of *Hesx1* leads to a complex and severe phenotype characterized by variable forebrain midline defects, severe microphthalmia, associated with developmental defects of the pituitary gland [26, 27]. However, the few individuals so far identified with a *HESX1* mutation display a less severe phenotype which differs among patients [26, 27].

Taken together, these observations further underscore the need for great caution when extrapolating the expected human phenotypic anomalies from the description of any mouse model. Among the various mechanisms which may account for such phenotypic differences between mice and the human patients [58, 59], the following ones, listed in table 4, concern especially developmental abnormalities. (1) Most importantly, the mutations introduced in mice are often not functionally equivalent to the disease-causing mutations identified in humans. Indeed, except mutational events like gene deletions or gene defects leading to severely truncated proteins, the other mutations identified in humans are likely to result only in a partial loss of function of the corresponding protein, whereas mutant mice generated by targeted disruption of the same gene have a complete loss of function of the orthologous protein. In this regard, it is important to underline that the identification of a sequence variation in a human candidate disease gene may require the development of sophisticated molecular and cellular tools to evaluate its functional consequences, i.e. to determine whether the sequence variation is indeed a disease-causing mutation and to elucidate the mechanism by which the function of the corresponding protein is altered. (2) In spite of the strong evolutionary conservation of coding sequences for developmentally important genes, significant human-mouse differences in embryonic expression patterns have already been documented for a variety of genes [60]. As the developmental expression pattern of most human pituitary-specific genes has not yet been studied, the existence of such species-specific differences during pituitary organogenesis cannot be excluded. (3) The presence in one or the other species of redundant biochemical pathways may result in species-specific phenotypic abnormalities. (4) It is well known that, depending on the genetic background of the mouse mutants, the same mutation may be either lethal or viable [61, 62]. In addition, such differences in the genetic background may contribute not only to the phenotypic differences documented between species, but also to those reported among individuals carrying the same gene defect. It is indeed clearly established that in most Mendelian disorders, the phenotypes are, in fact, complex traits which result from interactions of the major disease-causing gene with both modifying genetic factors and environmental factors [63–65]. Moreover, difficulties in extrapolating the phenotypic data from mouse models may reflect the existence of stochastic events during embryonic development. These epigenetic events might account for the phenotypic differences observed among inbred animals carrying the same gene defect, as illustrated by the study of mice carrying a targeted disruption of one or two *Hesx1* alleles [26]. Finally, the investigation of patients with mutations in pituitary-specific transcription factors like *PROP1* [57, 66] or *LHX3* [30] also revealed that both the pituitary morphology and the nature and severity of the hormonal deficit can dramatically vary with time in the same patient.

Conclusions

The molecular bases of pituitary defects are just beginning to be elucidated in humans. The identification of the genes underlying these heterogeneous conditions has so far relied on the analysis of mouse mutations resulting in analogous disease phenotypes. The candidate gene approach, therefore, undoubtedly represents a powerful system for deciphering the genetic causes of pituitary developmental defects in humans. The recent development of computational biology, a new interdisciplinary field accepting the challenge to create database systems capable of integrating knowledge from various biological and medical sources, simply underlines this point [67a]. In this regard, one should keep in mind that more than 1,000 mouse mutations are currently recorded in the Jackson Laboratory's Genome Database [67b]. Such information will become extremely valuable, relative to that expected from genetically engineered mice, these latter models (generated by means of quite sophisticated technologies) presenting specific limitations for extrapolation in humans. However, all such approaches must be based on accurate descriptions of the disease phenotype in the model used. Similarly, to facilitate identification of the molecular bases of pituitary developmental defects in humans, a most complete description of the disease phenotypes of the patients *and* their family members should be recorded, including not only pituitary-specific defects, but also all extrapituitary abnormalities, together with their intrafamilial mode of inheritance.

Nevertheless, the expected differences in the disease phenotype between that of human patients and that of murine models of pituitary defects leave open the possibility that only a systematic analysis of numerous candidate genes in large population samples obtained from patients presenting various pituitary phenotypes will be successful in identifying the underlying disease-causing genes. In fact, the studies that led to the discovery of the first *HESX1* and *LHX3* mutations were partly based on this approach [26, 30]. In addition, one cannot exclude the possibility that the pituitary abnormality of several patients may result from mutations at more than one locus. Indeed, in other highly heterogeneous disorders of genetic origin like retinitis pigmentosa [68], junctional epidermolysis bullosa [69] or nonsyndromic deafness [70], the identification of a single mutation in the heterozygous state in one gene may result in a mild or undiagnosed phenotype, whereas the occurrence of a second mutation in a different gene is required for determination of the genetic basis of these disorders. It is tempting to speculate that similar digenic patterns of inheritance may also account for abnormalities in pituitary development. Overall, such 'fishing' approaches for identification of the molecular bases of pituitary defects emphasize the need to perform these studies in the framework of large collaborative networks with the participation of physicians from different clinical centers who have these rare patients in charge, and biologists involved in the broad areas of embryological, cellular and genetic aspects of pituitary development in various species.

References

1 Couly G, Le Douarin NM: The fate map of the cephalic neural primordium at the presomitic to the 3-somite stage in the avian embryo. Development 1988;103:101–113.

2 Eagleson GW, Harris WA: Mapping of the presumptive brain regions in the neural plate of *Xenopus laevis.* J Neurobiol 1990;21:427–440.

3 Osumi-Yamashita N, Ninomiya Y, Doi H, Eto K: The contribution of both forebrain and midbrain crest cells to the mesenchyme in the frontonasal mass of mouse embryos. Dev Biol 1994;164:409–419.

4 Dubois PM, El Amraoui A: Embryology of the pituitary gland. Trends Endocrinol Metab 1995;6: 1–7.

5 Tremblay JJ, Lanctot C, Drouin J: The pan-pituitary activator of transcription, Ptx1 (pituitary homeobox 1), acts in synergy with SF-1 and Pit1 and is an upstream regulator of the Lim-homeodomain gene Lim3/Lhx3. Mol Endocrinol 1998;12:428–441.

6 Dasen JS, Rosenfeld MG: Combinatorial codes in signaling and synergy: Lessons from pituitary development. Curr Opin Genet Dev 1999;9:566–574.

7 Dasen JS, Rosenfeld MG: Signaling mechanisms in pituitary morphogenesis and cell fate determination. Curr Opin Cell Biol 1999;11:669–677.

8 Kioussi C, Carriere C, Rosenfeld MG: A model for the development of the hypothalamic-pituitary axis: Transcribing the hypophysis. Mech Dev 1999;81:23–35.

9 Watkins-Chow DE, Camper SA: How many homeobox genes does it take to make a pituitary gland? Trends Genet 1998;14:284–290.

10 Burrows H, Douglas K, Seasholtz A, Camper S: Genealogy of anterior pituitary gland: Tracing a family tree. Trends Endocrinol Metab 1999;10:343–352.

11 Sheng HZ, Westphal H: Early steps in pituitary organogenesis. Trends Genet 1999;15:236–240.

12 Li S, Crenshaw EBd, Rawson EJ, Simmons DM, Swanson LW, Rosenfeld MG: Dwarf locus mutants lacking three pituitary cell types result from mutations in the POU-domain gene pit-1. Nature 1990;347:528–533.

13 Andersen B, Rosenfeld MG: Pit-1 determines cell types during development of the anterior pituitary gland. A model for transcriptional regulation of cell phenotypes in mammalian organogenesis. J Biol Chem 1994;269:29335–29338.

14 Andersen B, Pearse RV 2nd, Jenne K, Sornson M, Lin SC, Bartke A, Rosenfeld MG: The Ames dwarf gene is required for Pit-1 gene activation. Dev Biol 1995;172:495–503.

15 Sornson MW, et al: Pituitary lineage determination by the Prophet of Pit-1 homeodomain factor defective in Ames dwarfism. Nature 1996;384:327–333.

16 Pfaffle RW, et al: Mutation of the POU-specific domain of Pit-1 and hypopituitarism without pituitary hypoplasia. Science 1992;257:1118–1121.

17 Radovick S, Nations M, Du Y, Berg LA, Weintraub BD, Wondisford FE: A mutation in the POU-homeodomain of Pit-1 responsible for combined pituitary hormone deficiency. Science 1992;257: 1115–1118.

18 Tatsumi K, Miyai K, Notomi T, Kaibe K, Amino N, Mizuno Y, Kohno H: Cretinism with combined hormone deficiency caused by a mutation in the PIT1 gene. Nat Genet 1992;1:56–58.

19 Wu W, et al: Mutations in PROP1 cause familial combined pituitary hormone deficiency. Nat Genet 1998;18:147–149.

20 Duquesnoy P, et al: Human Prop-1: Cloning, mapping, genomic structure. Mutations in familial combined pituitary hormone deficiency. FEBS Lett 1998;437:216–220.

21 Wajnrajch MP, Gertner JM, Harbison MD, Chua SC Jr, Leibel RL: Nonsense mutation in the human growth hormone-releasing hormone receptor causes growth failure analogous to the little (lit) mouse. Nat Genet 1996;12:88–90.

22 Netchine I, Talon P, Dastot F, Vitaux F, Goossens M, Amselem S: Extensive phenotypic analysis of a family with growth hormone (GH) deficiency caused by a mutation in the GH-releasing hormone receptor gene. J Clin Endocrinol Metab 1998;83:432–436.

23 Maheshwari HG, Silverman BL, Dupuis J, Baumann G: Phenotype and genetic analysis of a syndrome caused by an inactivating mutation in the growth hormone-releasing hormone receptor: Dwarfism of Sindh. J Clin Endocrinol Metab 1998;83:4065–4074.

24 Lin SC, Lin CR, Gukovsky I, Lusis AJ, Sawchenko PE, Rosenfeld MG: Molecular basis of the little mouse phenotype and implications for cell type-specific growth. Nature 1993;364:208–213.
25 Godfrey P, Rahal JO, Beamer WG, Copeland NG, Jenkins NA, Mayo KE: GHRH receptor of little mice contains a missense mutation in the extracellular domain that disrupts receptor function. Nat Genet 1993;4:227–232.
26 Dattani MT, et al: Mutations in the homeobox gene HESX1/Hesx1 associated with septo-optic dysplasia in human and mouse. Nat Genet 1998;19:125–133.
27 Thomas PQ, et al: Heterozygous HESX1 mutations associated with isolated congenital pituitary hypoplasia and septo-optic dysplasia. Hum Mol Genet 2001;10:39–45.
28 Sheng HZ, et al: Specification of pituitary cell lineages by the LIM homeobox gene Lhx3. Science 1996;272:1004–1007.
29 Sheng HZ, Moriyama K, Yamashita T, Li H, Potter SS, Mahon KA, Westphal H: Multistep control of pituitary organogenesis. Science 1997;278:1809–1812.
30 Netchine I, et al: Mutations in LHX3 result in a new syndrome revealed by combined pituitary hormone deficiency. Nat Genet 2000;25:182–186.
31 Li H, Zeitler PS, Valerius MT, Small K, Potter SS: Gsh-1, an orphan Hox gene, is required for normal pituitary development. EMBO J 1996;15:714–724.
32 Kimura S, Hara Y, Pineau T, Fernandez-Salguero P, Fox CH, Ward JM, Gonzalez FJ: The T/ebp null mouse: Thyroid-specific enhancer-binding protein is essential for the organogenesis of the thyroid, lung, ventral forebrain, and pituitary. Genes Dev 1996;10:60–69.
33 Takuma N, et al: Formation of Rathke's pouch requires dual induction from the diencephalon. Development 1998;125:4835–4840.
34 Topilko P, Schneider-Maunoury S, Levi G, Trembleau A, Gourdji D, Driancourt MA, Rao CV, Charnay P: Multiple pituitary and ovarian defects in Krox-24 (NGFI-A, Egr-1)-targeted mice. Mol Endocrinol 1998;12:107–122.
35 Acampora D, et al: Transient dwarfism and hypogonadism in mice lacking Otx1 reveal prepubescent stage-specific control of pituitary levels of GH, FSH and LH. Development 1998;125:1229–1239.
36 Jean D, Bernier G, Gruss P: Six6 (Optx2) is a novel murine Six3-related homeobox gene that demarcates the presumptive pituitary/hypothalamic axis and the ventral optic stalk. Mech Dev 1999;84:31–40.
37 Ericson J, Norlin S, Jessell TM, Edlund T: Integrated FGF and BMP signaling controls the progression of progenitor cell differentiation and the emergence of pattern in the embryonic anterior pituitary. Development 1998;125:1005–1015.
38 Treier M, Gleiberman AS, O'Connell SM, Szeto DP, McMahon JA, McMahon AP, Rosenfeld MG: Multistep signaling requirements for pituitary organogenesis in vivo. Genes Dev 1998;12:1691–1704.
39 Hu RM, et al: Gene expression profiling in the human hypothalamus pituitary adrenal axis and full-length cDNA cloning. Proc Natl Acad Sci USA 2000;97:9543–9548.
40 Pellegrini I, Barlier A, Gunz G, Figarella-Branger D, Enjalbert A, Grisoli F, Jaquet P: Pit-1 gene expression in the human pituitary and pituitary adenomas. J Clin Endocrinol Metab 1994;79:189–196.
41 Seidah NG, Barale JC, Marcinkiewicz M, Mattei MG, Day R, Chretien M: The mouse homeoprotein mLIM-3 is expressed early in cells derived from the neuroepithelium and persists in adult pituitary. DNA Cell Biol 1994;13:1163–1180.
42a Bach I: The LIM domain: Regulation by association. Mech Dev 2000;91:5–17.
42b Lamolet B, Puichino A-M, Lamonerie T, Gauthier Y, Brue T, Enjalbert A, Drouin J: A pituitary cell-restricted T box factor, Tpit, activates POMC transcription in cooperation with Pitx homeoproteins. Cell 2001;104:849–859.
43 Collins FS: Shattuck lecture – Medical and societal consequences of the Human Genome Project. N Engl J Med 1999;341:28–37.
44 Larizza D, Maraschio P, Maghnie M, Sampaolo P: Hypogonadism in a patient with balanced X/18 translocation and pituitary hormone deficiency. Eur J Pediatr 1993;152:424–427.

45 Siegel SF, Ahdab-Barmada M, Arslanian S, Foley TP Jr: Ectopic posterior pituitary tissue and paracentric inversion of the short arm of chromosome 1 in twins. Eur J Endocrinol 1995;133:87–92.
46 Lagerstrom-Fermer M, et al: X-linked recessive panhypopituitarism associated with a regional duplication in Xq25-q26. Am J Hum Genet 1997;60:910–916.
47 Hol FA, et al: Identification and characterization of an Xq26-q27 duplication in a family with spina bifida and panhypopituitarism suggests the involvement of two distinct genes. Genomics 2000;69: 174–181.
48 Lander ES, Botstein D: Homozygosity mapping: A way to map human recessive traits with the DNA of inbred children. Science 1987;236:1567–1570.
49 Miano MG, et al: Pitfalls in homozygosity mapping. Am J Hum Genet 2000;67:1348–1351.
50a Strausberg RL, Feingold EA, Klausner RD, Collins FS: The mammalian gene collection. Science 1999;286:455–457.
50b Lander ES, et al: Initial sequencing and analysis of the human genome. Nature 2001;409:860–921.
50c McPherson JD, et al: A physical map of the human genome. Nature 2001;409:934–941.
50d Venter JC, et al: The sequence of the human genome. Science 2001;291:1304–1351.
50e Caron H, et al: The human transcriptome map: Clustering of highly expressed genes in chromosomal domains. Science 2001;291:1289–1292.
51 Kruglyak L: The use of a genetic map of biallelic markers in linkage studies. Nat Genet 1997;17: 21–24.
52a Kruglyak L: Prospects for whole-genome linkage disequilibrium mapping of common disease genes. Nat Genet 1999;22:139–144.
52b Sachidanandam R, et al: A map of human genome sequence variation containing 1.42 million single nucleotide polymorphisms. Nature 2001;409:928–933.
53 O'Brien SJ, et al: The promise of comparative genomics in mammals. Science 1999;286:458–462, 479–481.
54 Hacia JG, et al: Determination of ancestral alleles for human single-nucleotide polymorphisms using high-density oligonucleotide arrays. Nat Genet 1999;22:164–167.
55 Gu H, Marth JD, Orban PC, Mossmann H, Rajewsky K: Deletion of a DNA polymerase beta gene segment in T cells using cell type-specific gene targeting. Science 1994;265:103–106.
56 Chambers CA: TKO'ed: Lox, stock and barrel. Bioessays 1994;16:865–868.
57 Mendonca BB, Osorio MG, Latronico AC, Estefan V, Lo LS, Arnhold IJ: Longitudinal hormonal and pituitary imaging changes in two females with combined pituitary hormone deficiency due to deletion of A301,G302 in the PROP1 gene. J Clin Endocrinol Metab 1999;84:942–945.
58 Wynshaw-Boris A: Model mice and human disease. Nat Genet 1996;13:259–260.
59 Erickson RP: Why isn't a mouse more like a man? Trends Genet 1989;5:1–3.
60 Fougerousse F, et al: Human-mouse differences in the embryonic expression patterns of developmental control genes and disease genes [published erratum appears in Hum Mol Genet 2000;9:659]. Hum Mol Genet 2000;9:165–173.
61 Liu JP, Baker J, Perkins AS, Robertson EJ, Efstratiadis A: Mice carrying null mutations of the genes encoding insulin-like growth factor I (Igf-1) and type 1 IGF receptor (Igf1r). Cell 1993;75:59–72.
62 Threadgill DW, et al: Targeted disruption of mouse EGF receptor: Effect of genetic background on mutant phenotype. Science 1995;269:230–234.
63 Romeo G, McKusick VA: Phenotypic diversity, allelic series and modifier genes. Nat Genet 1994;7: 451–453.
64 Dipple KM, McCabe ER: Phenotypes of patients with 'simple' Mendelian disorders are complex traits: Thresholds, modifiers and systems dynamics. Am J Hum Genet 2000;66:1729–1735.
65 Cazeneuve C, et al: Identification of MEFV-independent modifying genetic factors for familial Mediterranean fever. Am J Hum Genet 2000;67:1136–1143.
66 Fluck C, Deladoey J, Rutishauser K, Eble A, Marti U, Wu W, Mullis PE: Phenotypic variability in familial combined pituitary hormone deficiency caused by a PROP1 gene mutation resulting in the substitution of Arg → Cys at codon 120 (R120C). J Clin Endocrinol Metab 1998;83:3727–3734.
67a Blake JA, Richardson JE, Davisson MT, Eppig JT: The Mouse Genome Database (MGD): Genetic and genomic information about the laboratory mouse. The Mouse Genome Database Group. Nucleic Acids Res 1999;27:95–98.

67b Roos DS: Computational biology. Bioinformatics – trying to swim in a sea of data. Science 2001; 291:1260–1261.
68 Kajiwara K, Berson EL, Dryja TP: Digenic retinitis pigmentosa due to mutations at the unlinked peripherin/RDS and ROM1 loci. Science 1994;264:1604–1608.
69 Floeth M, Bruckner-Tuderman L: Digenic junctional epidermolysis bullosa: Mutations in COL17A1 and LAMB3 genes. Am J Hum Genet 1999;65:1530–1537.
70 Balciuniene J, Dahl N, Borg E, Samuelsson E, Koisti MJ, Pettersson U, Jazin EE: Evidence for digenic inheritance of nonsyndromic hereditary hearing loss in a Swedish family. Am J Hum Genet 1998;63:786–793.

Serge Amselem, Institut National de la Santé et de la Recherche Médicale (Unité 468)
Hôpital Henri-Mondor, F–94010 Créteil (France)
Tel. +33 1 49 81 28 73, Fax +33 1 49 81 28 42, E-Mail serge.amselem@im3.inserm.fr

Rappaport R, Amselem S (eds): Hypothalamic-Pituitary Development. Genetic and Clinical Aspects.
Endocr Dev. Basel, Karger, 2001, vol 4, pp 48–60

Growth Hormone-Releasing Hormone Receptors and Pituitary Development

Gerhard Baumann

Center for Endocrinology, Metabolism and Molecular Medicine, Department of Medicine, Northwestern University Medical School, Chicago, Ill., USA

Growth hormone-releasing hormone (GHRH) is the principal hypophysiotropic factor stimulating growth hormone (GH) production. In addition, it is an important player in pituitary organogenesis. Recent evidence derived from inactivating mutations in the GHRH receptor (GHRH-R) has highlighted the crucial importance of GHRH signaling for pituitary development, GH secretion and somatic growth. This chapter summarizes the present state of knowledge of the GHRH system as it pertains to pituitary development and function.

The GHRH/GHRH-R System

GHRH

GHRH is a 40–44-amino acid linear peptide synthesized and secreted in the hypothalamus, as well as, to a lesser extent, in other areas of the central nervous system, the gut and the endocrine pancreas [1]. It is structurally related to other neuropeptides, such as secretin, vasoactive intestinal peptide (VIP), pituitary adenylate cyclase-activating peptide (PACAP), glucagon and gastric inhibitory peptide (GIP) [2]. GHRH is released into the pituitary portal circulation and binds to GHRH-Rs on somatotropes [3] in the anterior pituitary gland, where it promotes GH synthesis and GH release from secretory granules. In addition to these acute effects, GHRH is critical for normal somatotrope development during the later stages of pituitary ontogeny (see below). In rare instances, GHRH overproduction, usually from an ectopic source (e.g., a carcinoid), can lead to somatotrope hyperplasia, GH overproduction, and acromegaly [4]. Indeed, GHRH was first isolated from such an ectopic source, islet cell tumors [5, 6].

GHRH is released in a pulsatile manner into the pituitary portal circulation; it is the main pacesetter for pulsatile GH secretion that is observed in virtually all species examined. The effect of GHRH on GH secretion is modulated by the prevailing somatostatin level/tone, which acts in an inhibitory manner dampening GH secretion. A third factor involved in GH regulation, whose physiological role is still poorly understood, is the recently discovered Ghrelin [7], a natural analogue of the GH secretagogues that have been investigated for a number of years [8]. Ghrelin is a 28-amino acid acylated peptide expressed primarily in the stomach, but also in the hypothalamus [7]. There is a strong but poorly understood synergistic interaction between the GHRH and the GH secretagogue systems [9].

GHRH is derived from a precursor molecule, pro-GHRH, the product of a single gene on chromosome 20p in humans [10, 11]. The GHRH gene is expressed in several extracerebral tissues, including the gonad [1]. In most instances, the biological role of GHRH in extrahypothalamic/extrapituitary tissues is unknown. In the brain, it may be involved in sleep and feeding regulation. An intriguing potential role for GHRH gene expression has been proposed in the gonad, where pro-GHRH may be processed differentially to yield a carboxy-terminal peptide distinct from GHRH (GHRH-related peptide or GHRH-rP) [12]. GHRH-rP has been postulated to play a role in Sertoli cell function and spermatogenesis [12], and a recent report of a transgenic mouse overexpressing GHRH-rP suggests that it may also be involved in hematopoiesis [13]. GHRH is rapidly inactivated in the circulation by an aminodipeptidase [14]. GHRH levels in the peripheral circulation are of limited usefulness because of this instability and because they do not reflect pituitary portal GHRH. An exception is the ectopic GHRH syndrome, where peripheral plasma GHRH levels are very high and pathognomonic [4].

Abnormalities in GHRH production or structure have long been postulated as a potential basis for idiopathic, isolated GH deficiency, particularly in familial cases. Surprisingly, to date no evidence has been obtained for a mutant GHRH gene. Thus, in contrast to overproduction of GHRH from ectopic or eutopic sources as a cause of acromegaly, GHRH underproduction (or a bioinactive GHRH) as the basis of GH deficiency remains to be discovered.

The GHRH-R

The GHRH-R was first cloned by three groups in 1992 [15–17]. It belongs to family B of the seven transmembrane domain, G-protein-coupled receptors [18]. It shares sequence homology with receptors for VIP, secretin, PACAP, parathyroid hormone, calcitonin and corticotropin-releasing hormone [19]. It is evident from this list of receptors and that of the corresponding ligands (see above) that ligands and receptors have co-evolved in parallel. The mature GHRH-R is a single chain, 401-amino acid glycoprotein, with a long extracellular domain, seven

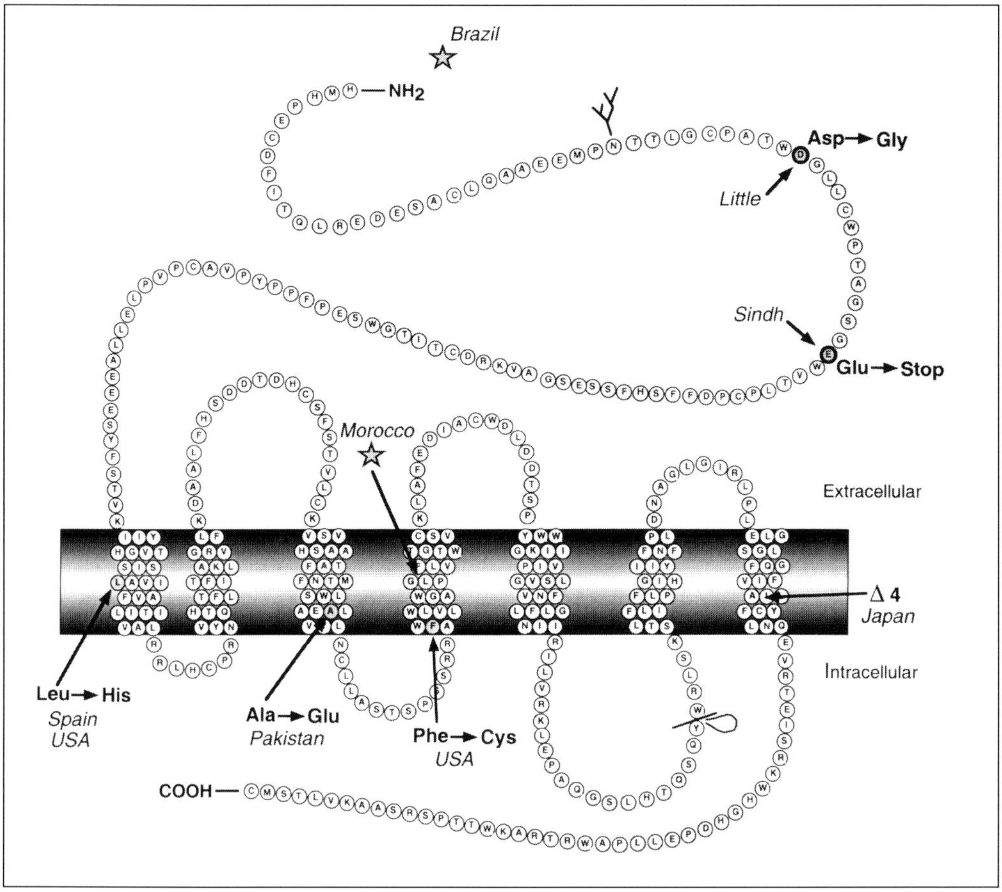

Fig. 1. Primary structure of the mature human GHRH-R. The branched tree structure denotes glycosylation. The locations of the described mutations are indicated, with the origin of the affected patients ('Little' denotes the little mouse). Stars denote splice site mutations, D4 a 4-base deletion. The Brazilian mutation truncates the signal peptide upstream from the mature receptor (indicated by the star). The site in the third intracellular loop, where receptor variants are truncated (human) or contain an insertion (rodents), is indicated by a bar or small loop, respectively.

transmembrane helices connected by loops, and a 43-amino acid intracellular domain (fig. 1). The sequence is preceded by a 22-residue signal peptide that is cleaved off during translation. The extracellular domain contains a single glycosylation site at Asn^{28} and 7 cysteines, most of which are probably disulfide-linked. Following activation by ligand binding, the GHRH-R couples with G_s protein and signals through the cAMP/protein kinase A pathway. Less well characterized is the pathway leading to the rapid Ca^{2+} influx that follows GHRH binding to its receptor; inositol triphosphate and protein kinase C pathways have been implicated. GHRH binding is complex and involves not only the extracellular domain, but also portions of the extracellular loops and transmembrane helices [20].

In humans, the GHRH-R gene is located on the short arm of chromosome 7, in the mouse it resides on chromosome 6. The human GHRH-R gene spans 15.5 kb in its coding region and consists of 13 exons and 12 introns [21, 22]. The genomic organization is similar in rodents, although the introns are shorter [23]. Rodents, in addition to the main GHRH-R, express a splice variant containing a 41-amino acid insertion in the third intracellular loop [15, 16]. This variant, which is expressed only in small quantities, is nonfunctional with respect to cAMP generation [23]; its biological significance is not clear. In humans, splice variants truncated in the third intracellular loop have been described in normal and adenomatous pituitary [24, 25]. These variants are also predicted to be nonfunctional [24, 25].

The GHRH-R is expressed primarily in the pituitary gland, but also to a substantial degree in the kidney [26]. Lesser amounts are expressed in virtually all other tissues, although it is not clear whether these levels are sufficiently high to be functionally competent [27]. Important information about GHRH-R function in the whole organism has been obtained from genetic GHRH-R deficiency states (see below); they have provided no evidence for a critical function of the GHRH-R in extrapituitary or extraneural tissues. In the pituitary gland, GHRH-R expression is dependent on the transcription factor Pit-1 [28, 29]. Pit-1 expression precedes that of the GHRH-R during pituitary ontogeny; it leads to upregulation of GHRH-R synthesis in somatotropes, which in turn permits GHRH-dependent somatotrope proliferation (see below). Glucocorticoids, as well as GHRH itself, are known to regulate GHRH-R expression. While glucocorticoids stimulate GHRH-R expression [28, 29], GHRH has been reported to either stimulate or inhibit GHRH-R synthesis [30–32]. The promoter region of GHRH-R gene contains response elements for Pit-1 and glucocorticoids [22, 23, 33].

GHRH-R Deficiency

Causes and Clinical Consequences

Considerable new insight into the biology of GHRH and its receptor has been gained from naturally occurring inactivating mutations in the GHRH-R gene. The first to be described was that in the little (*lit/lit*) mouse. This mouse was first recognized as a dwarfed strain in the Jackson laboratories in the mid-1970s [34]. The cause of dwarfism was identified as GH deficiency, and subsequently

Table 1. Phenotypic characteristics in genetic GHRH-R deficiency

	Little mouse	Humans
Proportionate dwarfism	Yes	Yes
Postnatal growth failure	Yes	Yes
Microcephaly	Yes	Yes
Dysmorphism	No	No or minimal
Pituitary hypoplasia	Yes	Yes
Delayed puberty	Yes	Yes
Fertility	Yes	Yes
Lactation	± normal	Normal
Isolated GH deficiency	Yes	Yes
Abnormal body composition	Yes	Not known
Intelligence	?	Appears normal
Hypoglycemia	?	No
Micropenis	?	No

the little mouse was found to be unresponsive to GHRH [35, 36]. The defect is transmitted in an autosomal recessive manner. Linkage analysis implicated a locus near the GHRH-R gene on chromosome 6, and sequencing revealed a missense mutation in the extracellular domain of the GHRH-R that converts a highly conserved aspartate residue (Asp^{60}) to glycine [37, 38]. The mutant GHRH-R, although normally processed and displayed in the plasma membrane, is dysfunctional because its is unable to bind ligand [39]. The consequence is resistance to GHRH.

Salient features of the little mouse are isolated GH deficiency, postnatal growth retardation, proportionate dwarfism, abnormal body composition consistent with GH deficiency, delayed puberty, pituitary hypoplasia and microcephaly (table 1) [34, 40]. Most of the phenotypic abnormalities can be attributed to GH deficiency, with the exception of pituitary hypoplasia (discussed below) and microcephaly. The latter has been attributed to abnormal glial and neuronal development in the central nervous system, with a smaller brain resulting in decreased cranial size [41, 42]. Despite these findings, the little mouse exhibits no obvious neurological or behavioral abnormality. Microcephaly may be a direct consequence of GHRH resistance in the brain (GHRH being a neurotransmitter), rather than secondary to GH deficiency. The little mouse shows normal fertility; it has some difficulty with lactation with the first, but not with later pregnancies [34]. Treatment of primiparas with GH or prolactin corrects the lactational limitation [43].

Recently, the human counterpart of the little mouse has been identified. Wajnrajch et al. [44] first reported a mutation in the GHRH-R as the basis for GH deficiency in 2 patients of Indian descent in New York. Maheshwari et al. [45, 46] independently described the same GHRH-R mutation in a large consanguineous kindred affected by a high prevalence of severe dwarfism in Pakistan (Dwarfs of Sindh). Netchine et al. [47] reported two GH-deficient patients of Sri Lankan origin with the same mutation in Paris. It is of interest that in all three instances the families reside in or originate from the Indian subcontinent. Preliminary studies indicate that they are related, albeit distantly [48]. A founder mutation may have occurred several hundred years ago [48], although in that case it is not clear why the dwarf phenotype has not emerged or been noticed earlier. The mutation in all three families is a nonsense mutation (GAG $\rightarrow$ TAG, E72X or E50X, depending on whether or not the numbering includes the signal peptide) in the extracellular domain that severely truncates the GHRH-R (fig. 1). The mutant receptor lacks not only a ligand-binding site but also most of its backbone, including the transmembrane domains, intracellular and extracellular loops, and cytoplasmic domain. It is likely that the mutant GHRH-R is not even expressed but rapidly degraded in the endoplasmic reticulum. The dwarf phenotype is transmitted in a autosomal recessive mode. Thus, in the homozygous state, this mutation confers severe GHRH resistance and GH deficiency. In the heterozygous state, the mutation has few consequences on physical phenotype, although mild to moderate biochemical abnormalities in the GH-IGF axis are present [45, 46]. More recently, another point mutation in the GHRH-R has been described as the basis of GH-deficient dwarfism in a large, inbred Brazilian kindred [49]. In that case, the mutation affects the splice donor site of intron 1 ($g^{+1} \rightarrow a$) with presumed readthrough into intron 1 and premature truncation of the GHRH-R within an abnormal signal peptide. The functional consequence is predictably similar to that of the Sindh mutation described above. Indeed, the Brazilian patients also have severe GHRH resistance, with a phenotype that is very similar to that of the Pakistani/Indian patients [50, 51]. Several additional mutations have recently been identified in cases of isolated GH deficiency. They include missense mutations in transmembrane helices 1, 3 and 4 [52], a 4-base deletion in transmembrane helix 7 disrupting that helix and causing absence of the intracellular domain [53], and a splice donor site mutation of intron 7 ($g^{+1} \rightarrow c$) resulting in truncation within transmembrane helix 4 [54]. The inactivating mutations reported to date are summarized in figure 1. It appears likely that additional mutations will be discovered.

The principal features of human GHRH-R deficiency are listed in table 1. They are remarkably congruous with those found in the little mouse, but extend the information available about the consequences of GHRH resistance. Most of the anomalies, such as proportionate dwarfism, postnatal growth retardation, delayed puberty and the biochemical findings, can be ascribed to isolated GH deficiency. As in the little mouse, pituitary hypoplasia and relative microcephaly cannot be explained by GH deficiency alone, but likely represent the absence of direct GHRH effects. Patients with GHRH-R deficiency have much smaller heads (4 standard deviations (SD) below the norm) [46] than patients with GH deficiency resulting from other causes (0–2 SD below the norm) [55–57]. In both cases, body size is 6–8 SD below the norm. This relative parallelism between head and body size reduction in GHRH-R deficiency accounts for the near-normal, but miniaturized aspect of the affected patients [46]. Intelligence and neurological function is normal despite the small head size. Fertility and lactation also appear normal. In view of the widespread expression of the GHRH-R among tissues, the absence of any other discernible phenotypic aberrations is notable. Thus, it appears that the GHRH-R subserves no critical function in extrapituitary and extraneural tissues. Of interest, patients with GHRH-R deficiency fail to secrete GH in response to hexarelin, although the ACTH and prolactin responses to that GH secretagogue are intact [58]. This further corroborates that an intact GHRH signaling system is required for the ghrelin-type secretagogues to effect GH release, despite the fact that they interact with their own receptor [8].

Impact on Pituitary Development

Pituitary ontogeny depends on the sequential expression of several transcription factors, including Rpx, Ptx-1, Otx-1, Lhx-3, Prop-1 and Pit-1 [59]. These factors direct organ formation and differentiation of distinct cell lineages, including the somatotrope line. Terminal differentiation and the expansion of somatotropes late in pituitary development is governed by GHRH [38, 59]. In the absence of a functional GHRH system, somatotrope proliferation fails to take place. Since somatotropes, the dominant pituitary cell type, normally account for about half the pituitary volume, somatotrope hypoplasia results in a significantly reduced pituitary size. Relative lack of somatotropes is the reason for the pituitary hypoplasia seen in the little mouse, and likely explains the pituitary hypoplasia in humans with GHRH-R deficiency. Indeed, the pituitary volume in the latter is estimated at about 35–50% normal, based on magnetic resonance imaging [60]. The typical anatomy consists of a small anterior lobe flattened against the sella floor, with a normal stalk and a posterior lobe that is normal in size and position [47, 60]. Quantitative morphometry obtained in the little mouse shows a 75% reduction in somatotropes [61]. In addition, the GH content per remaining somatotrope is reduced by 60% compared to normal, resulting in a pituitary GH content of only about 10% of normal. Furthermore, the residual GH is not releasable with GHRH, as has been shown with isolated somatotropes in vitro [35].

Thus, GHRH-R deficiency or dysfunction results in a triple and possibly quadruple defect in the GH axis: First, somatotrope number is severely curtailed.

Second, the remaining somatotropes are unable to synthesize GH normally, which further reduces pituitary GH content. Third, the small amount of stored GH cannot be released by GHRH, the final common pathway for most hypothalamic secretory signals. Fourth, the putative (though still unproven) role of ghrelin in physiological GH secretion would also appear to be disabled in the absence of a normal GHRH signaling system. The result of these multiple limitations is one of the most severe types of GH-deficient dwarfism described to date, as judged by the degree of growth failure and the serum IGF-1 level [46, 49, 50].

Potential Role of the GHRH-R in Other Disease States

GHRH-R and Acromegaly

The converse of somatotrope hypoplasia and GH deficiency is somatotrope hyperplasia/neoplasia and GH excess, which causes acromegaly or gigantism in children. From the foregoing, it is reasonable to postulate that a hyperactive GHRH system may lead to GH hypersecretion and acromegaly. GHRH, when present in excess, clearly causes somatotrope hyperplasia, as best exemplified in the ectopic GHRH syndrome associated with acromegaly [62]. Transgenic mice overexpressing GHRH also develop pituitary hyperplasia [63] as well as somatotrope and somatomammotrope tumors [64]. It has been speculated that activating mutations in the GHRH-R may underlie some cases of acromegaly due to constitutive activation of the cAMP pathway, akin to those seen with activating mutations in the G_s protein [65, 66]. However, with two possible exceptions [67, 68], several extensive studies have failed to identify such activating mutations in the GHRH-R. An amino acid change in the extracellular domain of the GHRH-R ($Ala^{57} \rightarrow Thr$) found in two GH-secreting adenomas turned out to be a polymorphism, with the threonine-containing allele showing an enhanced cAMP response to GHRH, but normal basal activity [68]. Another amino acid change in the intracellular domain, found in two somatotrophic tumors, was associated with increased cAMP accumulation both in the basal and GHRH-stimulated state [67]. The potential role of these GHRH-R variants in tumorigenesis and acromegaly remains to be determined.

Splice Variants of the GHRH-R Gene and Cancer

Based on the observation that antagonists to GHRH can inhibit proliferation of certain malignancies not only in vivo, but also in vitro, Schally and Varga [69] postulated a direct effect of such antagonists on tumor cells and conducted a search for specific receptors for GHRH antagonists. Recent studies have identified such receptors on human renal carcinoma cells, with affinities that are at least 10-fold higher for GHRH antagonists than for GHRH itself [70], and further investigations identified four new GHRH-R splice variants in human cancer cell lines [71]. All four variants lack the N-terminal portion encoded by exons 1–3, which is replaced by a sequence derived from intron 3 that may encode an alternative signal peptide that precedes the exon 4 encoded sequence. Thus, the extracellular domain is shortened from the normal 108 residues to 40 of its most carboxy-terminal residues. In one variant, this shortened extracellular domain is followed by an otherwise normal downstream GHRH-R sequence; that variant may function as a cell-membrane-bound receptor. The others either lack critical regions because of skipped exons/frameshifts or, in one case, encode an entirely different amino acid sequence; their functionality as cell surface receptors is doubtful. The relationship of these GHRH-R gene splice variants to the abovementioned receptors for GHRH antagonists remains to be clarified. It is also unclear what the natural ligand for a 'GHRH antagonist receptor' might be. Nevertheless, these intriguing findings raise the possibility that the GHRH-R gene may be involved in more than classical GHRH action.

Conclusion

The GHRH-R is a member of family B of the G protein-coupled, seven transmembrane domain receptors. It confers GHRH responsivity to cells that express it and signals primarily through the cAMP pathway. Its principal site of expression is the somatotrope, where it is essential for cellular proliferation during pituitary development, GH synthesis and GH release. Another functionally important site of expression is in the central nervous system, where GHRH promotes glial and neuronal development. The recent recognition of genetic GHRH-R deficiency states has helped define the tissues where GHRH action is critical. GHRH-R deficiency leads to global failure of the GHRH-GH-IGF axis, with severe postnatal growth retardation and other characteristics of isolated GH deficiency. In addition, it leads to relative microcephaly through subnormal brain development, but that is not accompanied by detectable brain dysfunction. Although the GHRH-R is widely distributed among tissues, its deficiency has little direct consequence on organ function outside the pituitary and central nervous system. The possible existence of activating mutations in the GHRH-R as the basis for somatotropic tumors and acromegaly has been extensively sought but found to be rare. Preliminary data have identified two GHRH-R polymorphic variants that show enhanced signaling activity, although their role in tumorigenesis remains unclear. Recent studies in human neoplastic cells have identified high affinity binding sites for GHRH antagonists and novel GHRH-R gene splice variants. Such variants possibly mediate the direct inhibitory effects of GHRH antagonists on malignant cell proliferation.

Acknowledgements

This work was supported in part by grants from the Genentech Foundation for Growth and Development, the Human Growth Foundation, the Department of Veterans Affairs, the National Institutes of Health (RR00048), and the Northwestern University Intramural Grant Program.

References

1 Frohman LA, Jansson JO: Growth hormone-releasing hormone. Endocr Rev 1986;7:223–253.
2 Campbell RM, Scanes CG: Evolution of the growth hormone-releasing factor (GRF) family of peptides. Growth Regul 1992;2:175–191.
3 Morel G, Gallego R, Boulanger L, Pintos E, Garcia-Caballero T, Gaudreau P: Restricted presence of the growth hormone-releasing hormone receptor to somatotropes in rat and human pituitaries. Neuroendocrinology 1999;70:128–136.
4 Faglia G, Arosio M, Bazzoni N: Ectopic acromegaly. Endocrinol Metab Clin North Am 1992;21: 575–595.
5 Rivier J, Spiess J, Thorner M, Vale W: Characterization of a growth hormone-releasing factor from a human pancreatic islet tumour. Nature 1982;300:276–278.
6 Guillemin R, Brazeau P, Bohlen P, Esch F, Ling N, Wehrenberg WB: Growth hormone-releasing factor from a human pancreatic tumor that caused acromegaly. Science 1982;218:585–587.
7 Kojima M, Hosoda H, Date Y, Nakazato M, Matsuo H, Kangawa K: Ghrelin is a growth hormone-releasing acylated peptide from stomach. Nature 1999;402:656–660.
8 Smith RG, Van der Ploeg LH, Howard AD, Feighner SD, Cheng K, Hickey GJ, Wyvratt MJ Jr, Fisher MH, Nargund RP, Patchett AA: Peptidomimetic regulation of growth hormone secretion. Endocr Rev 1997;18:621–645.
9 Bowers CY, Reynolds GA, Durham D, Barrera CM, Pezzoli SS, Thorner MO: Growth hormone (GH)-releasing peptide stimulates GH release in normal men and acts synergistically with GH-releasing hormone. J Clin Endocrinol Metab 1990;70:975–982.
10 Gubler U, Monahan JJ, Lomedico PT, Bhatt RS, Collier KJ, Hoffman BJ, Bohlen P, Esch F, Ling N, Zeytin F, Brazeau P, Poonian MS, Gage LP: Cloning and sequence analysis of cDNA for the precursor of human growth hormone-releasing factor, somatocrinin. Proc Natl Acad Sci USA 1983; 80:4311–4314.
11 Mayo KE, Cerelli GM, Lebo RV, Bruce BD, Rosenfeld MG, Evans RM: Gene encoding human growth hormone-releasing factor precursor: Structure, sequence, and chromosomal assignment. Proc Natl Acad Sci USA 1985;82:63–67.
12 Breyer PR, Rothrock JK, Beaudry N, Pescovitz OH: A novel peptide from the growth hormone-releasing hormone gene stimulates Sertoli cell activity. Endocrinology 1996;137:2159–2162.
13 Fang S, Steinmetz R, Walker-King D, Zeng P, Vogelweid C, Cooper S, Hangcoc G, Broxmeyer HE, Pescovitz OH: Development of a transgenic mouse that overexpresses a novel product of the growth hormone-releasing hormone gene. Endocrinology 2000;141:1377–1383.
14 Frohman LA, Downs TR, Williams TC, Heimer EP, Pan YC, Felix AM: Rapid enzymatic degradation of growth hormone-releasing hormone by plasma in vitro and in vivo to a biologically inactive product cleaved at the NH2 terminus. J Clin Invest 1986;78:906–913.
15 Lin C, Lin SC, Chang CP, Rosenfeld MG: Pit-1-dependent expression of the receptor for growth hormone-releasing factor mediates pituitary cell growth. Nature 1992;360:765–768.
16 Mayo KE: Molecular cloning and expression of a pituitary-specific receptor for growth hormone-releasing hormone. Mol Endocrinol 1992;6:1734–1744.
17 Gaylinn BD, Harrison JK, Zysk JR, Lyons CE, Lynch KR, Thorner MO: Molecular cloning and expression of a human anterior pituitary receptor for growth hormone-releasing hormone. Mol Endocrinol 1993;7:77–84.

18 Kolakowski LF Jr: GCRDb: A G-protein-coupled receptor database. Receptors Channels 1994;2: 1–7.
19 Segre G, Goldring SR: Receptors for secretin, calcitonin, parathyroid hormone (PTH)/PTH-related peptide, vasoactive intestinal peptide, glucagon-like peptide 1, growth hormone-releasing hormone, and glucagon belong to a newly discovered G-protein-linked receptor family. Trends Endocrinol Metab 1993;4:309–314.
20 DeAlmeida VI, Mayo KE: Identification of binding domains of the growth hormone-releasing hormone receptor by analysis of mutant and chimeric receptor proteins. Mol Endocrinol 1998;12: 750–765.
21 Maheshwari HG, Baumann G: Genomic organization and structure of the human growth hormone-releasing hormone receptor gene. Proc 79th Annual Meeting of the Endocrine Society, Minneapolis 1997, p 156.
22 Petersenn S, Rasch AC, Heyens M, Schulte HM: Structure and regulation of the human growth hormone-releasing hormone receptor gene. Mol Endocrinol 1998;12:233–247.
23 Miller TL, Godfrey PA, DeAlmeida VI, Mayo KE: The rat growth hormone-releasing hormone receptor gene: Structure, regulation and generation of receptor isoforms with different signaling properties. Endocrinology 1999;140:4152–4165.
24 Hashimoto K, Koga M, Motomura T, Kasayama S, Kouhara H, Ohnishi T, Arita N, Hayakawa T, Sato B, Kishimoto T: Identification of alternatively spliced messenger ribonucleic acid encoding truncated growth hormone-releasing hormone receptor in human pituitary adenomas. J Clin Endocrinol Metab 1995;80:2933–2939.
25 Tang J, Lagace G, Castagne J, Collu R: Identification of human growth hormone-releasing hormone receptor splicing variants. J Clin Endocrinol Metab 1995;80:2381–2387.
26 Mayo KE, Miller TL, DeAlmeida V, Zheng J, Godfrey PA: The growth hormone-releasing hormone receptor: Signal transduction, gene expression and physiological function in growth regulation. Ann NY Acad Sci 1996;805:184–203.
27 Matsubara S, Sato M, Mizobuchi M, Niimi M, Takahara J: Differential gene expression of growth hormone (GH)-releasing hormone (GRH) and GRH receptor in various rat tissues. Endocrinology 1995;136:4147–4150.
28 Seifert H, Perrin M, Rivier J, Vale W: Growth hormone-releasing factor binding sites in rat anterior pituitary membrane homogenates: Modulation by glucocorticoids. Endocrinology 1985;117:424–426.
29 Miller TL, Mayo KE: Glucocorticoids regulate pituitary growth hormone-releasing hormone receptor messenger ribonucleic acid expression. Endocrinology 1997;138:2458–2465.
30 Miki N, Ono M, Murata Y, Ohsaki E, Tamitsu K, Yamada M, Demura H: Regulation of pituitary growth hormone-releasing factor (GRF) receptor gene expression by GRF. Biochem Biophys Res Commun 1996;224:586–590.
31 Aleppo G, Moskal SF 2nd, De Grandis PA, Kineman RD, Frohman LA: Homologous down-regulation of growth hormone-releasing hormone receptor messenger ribonucleic acid levels. Endocrinology 1997;138:1058–1065.
32 Horikawa R, Hellmann P, Cella SG, Torsello A, Day RN, Müller EE, Thorner MO: Growth hormone-releasing factor (GRF) regulates expression of its own receptor. Endocrinology 1996;137: 2642–2645.
33 Iguchi G, Okimura Y, Takahashi T, Mizuno I, Fumoto M, Takahashi Y, Kaji H, Abe H, Chihara K: Cloning and characterization of the 5′-flanking region of the human growth hormone-releasing hormone receptor gene. J Biol Chem 1999;274:12108–12114.
34 Eicher EM, Beamer WG: Inherited ateliotic dwarfism in mice. Characteristics of the mutation, 'little', on chromosome 6. J Hered 1976;67:87–91.
35 Jansson JO, Downs TR, Beamer WG, Frohman LA: Receptor-associated resistance to growth hormone-releasing factor in dwarf 'little' mice. Science 1986;232:511–512.
36 Clark RG, Robinson IC: Effects of a fragment of human growth hormone-releasing factor in normal and 'Little' mice. J Endocrinol 1985;106:1–5.
37 Godfrey P, Rahal JO, Beamer WG, Copeland NG, Jenkins NA, Mayo KE: GHRH receptor of little mice contains a missense mutation in the extracellular domain that disrupts receptor function. Nat Genet 1993;4:227–232.

38 Lin SC, Lin CR, Gukovsky I, Lusis AJ, Sawchenko PE, Rosenfeld MG: Molecular basis of the little mouse phenotype and implications for cell type-specific growth. Nature 1993;364:208–213.
39 Gaylinn BD, DeAlmeida VI, Lyons CE Jr, Wu KC, Mayo KE, Thorner MO: The mutant growth hormone-releasing hormone (GHRH) receptor of the little mouse does not bind GHRH. Endocrinology 1999;140:5066–5074.
40 Donahue LR, Beamer WG: Growth hormone deficiency in 'little' mice results in aberrant body composition, reduced insulin-like growth factor-I and insulin-like growth factor-binding protein-3 (IGFBP-3), but does not affect IGFBP-2, -1 or -4. J Endocrinol 1993;136:91–104.
41 Noguchi T: Retarded cerebral growth of hormone-deficient mice. Comp Biochem Physiol C 1991; 98:239–248.
42 Morisawa K, Sugisaki T, Kanamatsu T, Aoki T, Noguchi T: Factors contributing to cerebral hypomyelination in the growth hormone-deficient little mouse. Neurochem Res 1989;14:173–177.
43 Keough EM, Wood BG: Mammary gland development during pregnancy in the dwarf mouse mutant, 'little'. Tissue Cell 1979;11:773–780.
44 Wajnrajch MP, Gertner JM, Harbison MD, Chua SC Jr, Leibel RL: Nonsense mutation in the human growth hormone-releasing hormone receptor causes growth failure analogous to the little (lit) mouse. Nat Genet 1996;12:88–90.
45 Baumann G, Maheshwari H: The dwarfs of Sindh: Severe growth hormone (GH) deficiency caused by a mutation in the GH-releasing hormone receptor gene. Acta Paediatr Suppl 1997;423:33–38.
46 Maheshwari HG, Silverman BL, Dupuis J, Baumann G: Phenotype and genetic analysis of a syndrome caused by an inactivating mutation in the growth hormone-releasing hormone receptor: Dwarfism of Sindh. J Clin Endocrinol Metab 1998;83:4065–4074.
47 Netchine I, Talon P, Dastot F, Vitaux F, Goossens M, Amselem S: Extensive phenotypic analysis of a family with growth hormone (GH) deficiency caused by a mutation in the GH-releasing hormone receptor gene. J Clin Endocrinol Metab 1998;83:432–436.
48 Wajnrajch MP, Gertner JM, Harbison MD, Netchine I, Maheshwari HG, Baumann G, Amselem S, Leibel RL: Haplotype analysis of three kindreds with an identical (E72X) mutation in the growth hormone-releasing hormone receptor gene. Proc 81st Annual Meeting of the Endocrine Society, San Diego 1999, p 342.
49 Salvatori R, Hayashida CY, Aguiar-Oliveira MH, Phillips JA 3rd, Souza AH, Gondo RG, Toledo SP, Conceicao MM, Prince M, Maheshwari HG, Baumann G, Levine MA: Familial dwarfism due to a novel mutation of the growth hormone-releasing hormone receptor gene. J Clin Endocrinol Metab 1999;84:917–923.
50 Aguiar-Oliveira MH, Gill MS, de A Barretto ES, Alcantara MR, Miraki-Moud F, Menezes CA, Souza AH, Martinelli CE, Pereira FA, Salvatori R, Levine MA, Shalet SM, Camacho-Hübner C, Clayton PE: Effect of severe growth hormone (GH) deficiency due to a mutation in the GH-releasing hormone receptor on insulin-like growth factors (IGFs), IGF-binding proteins, and ternary complex formation throughout life. J Clin Endocrinol Metab 1999;84:4118–4126.
51 Hayashida CY, Gondo RG, Ferrari C, Toledo SPA, Salvatori R, Levine MA, Ezabella MCL, Abelin N, Gianella-Neto D, Wajchenberg BL: Familial growth hormone deficiency with mutated GHRH receptor gene: Clinical and hormonal findings in homozygous and heterozygous individuals from Itabaianinha. Eur J Endocrinol 2000;142:557–563.
52 Salvatori R, Fan X, Phillips JAI, Espigares-Martin R, Martin de Lara I, Freeman KL, Plotnick L, Al-Ashwal A, Levine MA: Three new mutations in the gene for the growth hormone (GH)-releasing hormone receptor in familial isolated GH deficiency type IB. J Clin Endocrinol Metab 2001;86: 273–279.
53 Horikawa R, Fujita K, Nakajima R, Gaylinn B, Tanaka T: A novel growth hormone-releasing hormone (GHRH) receptor mutation as a cause for isolated GH deficiency in a Japanese boy with severe short stature. Proc 82nd Meeting Endocrine Society 2000, p 482.
54 Stokvis-Brantsma WH, Blankenstein O, Oostdijk W, Roelfsma F, Pfäffle R, Wit JM: Endocrine findings and growth response to combined GH and GnRH agonist treatment in two siblings with a novel mutation of the GHRH receptor. Hormone Res 2000;53(suppl 2):57.
55 Goodman HG, Grumbach MM, Kaplan SL: Growth and growth hormone. II. A comparison of isolated growth-hormone deficiency and multiple pituitary-hormone deficiencies in 35 patients with idiopathic hypopituitary dwarfism. N Engl J Med 1968;278:57–68.

56 Burt L, Kulin HE: Head circumference in children with short stature secondary to primary hypothyroidism. Pediatrics 1977;59:628–630.
57 Laron Z, Roitman A, Kauli R: Effect of human growth hormone therapy on head circumference in children with hypopituitarism. Clin Endocrinol (Oxf) 1979;10:393–399.
58 Maheshwari HG, Rahim A, Shalet SM, Baumann G: Selective lack of growth hormone (GH) response to the GH-releasing peptide hexarelin in patients with GH-releasing hormone receptor deficiency. J Clin Endocrinol Metab 1999;84:956–959.
59 Treier M, Gleiberman AS, O'Connell SM, Szeto DP, McMahon JA, McMahon AP, Rosenfeld MG: Multistep signaling requirements for pituitary organogenesis in vivo. Genes Dev 1998;12:1691–1704.
60 Murray RA, Maheshwari HG, Russell EJ, Baumann G: Pituitary hypoplasia in patients with a mutation in the growth hormone-releasing hormone receptor gene. AJNR Am J Neuroradiol 2000; 21:685–689.
61 Wilson DB, Wyatt DP, Gadler RM, Baker CA: Quantitative aspects of growth hormone cell maturation in the normal and little mutant mouse. Acta Anat 1988;131:150–155.
62 Sano T, Asa SL, Kovacs K: Growth hormone-releasing hormone-producing tumors: Clinical, biochemical, and morphological manifestations. Endocr Rev 1988;9:357–373.
63 Mayo KE, Hammer RE, Swanson LW, Brinster RL, Rosenfeld MG, Evans RM: Dramatic pituitary hyperplasia in transgenic mice expressing a human growth hormone-releasing factor gene. Mol Endocrinol 1988;2:606–612.
64 Asa SL, Kovacs K, Stefaneanu L, Horvath E, Billestrup N, Gonzalez-Manchon C, Vale W: Pituitary adenomas in mice transgenic for growth hormone-releasing hormone. Endocrinology 1992;131: 2083–2089.
65 Landis CA, Masters SB, Spada A, Pace AM, Bourne HR, Vallar L: GTPase inhibiting mutations activate the alpha chain of Gs and stimulate adenylyl cyclase in human pituitary tumours. Nature 1989;340:692–696.
66 Weinstein LS, Shenker A, Gejman PV, Merino MJ, Friedman E, Spiegel AM: Activating mutations of the stimulatory G protein in the McCune-Albright syndrome. N Engl J Med 1991;325:1688–1695.
67 Petersenn S, Schneider-Merck T, Luedecke D, Beil FU, Schulte HM: Mutations of the growth hormone-releasing hormone receptor in somatotrophic pituitary adenomas. Proc 82nd Meeting Endocrine Society 2000, p 162.
68 Adams EF, Symowski H, Buchfelder M, Poyner DR: A polymorphism in the growth hormone (GH)-releasing hormone (GHRH) receptor gene is associated with elevated response to GHRH by human pituitary somatotrophinomas in vitro. Biochem Biophys Res Commun 2000;275:33–36.
69 Schally AV, Varga JL: Antagonistic analogs of growth hormone-releasing hormone: New potential antitumor agents. Trends Endocrinol Metab 1999;10:383–391.
70 Halmos G, Schally AV, Varga JL, Plonowski A, Rekasi Z, Czompoly T: Human renal cell carcinoma expresses distinct binding sites for growth hormone-releasing hormone. Proc Natl Acad Sci USA 2000;97:10555–10560.
71 Rekasi Z, Czompoly T, Schally AV, Halmos G: Isolation and sequencing of cDNAs for splice variants of growth hormone-releasing hormone receptors from human cancers. Proc Natl Acad Sci USA 2000;97:10561–10566.

G. Baumann, MD, 303 East Chicago Avenue, Chicago, IL 60611 (USA)
Tel. +1 312 503 4128, Fax +1 312 908 9032, E-Mail gbaumann@northwestern.edu

Rappaport R, Amselem S (eds): Hypothalamic-Pituitary Development. Genetic and Clinical Aspects.
Endocr Dev. Basel, Karger, 2001, vol 4, pp 61–76

Pituitary Transcription Factors, POU1F1 and PROP1 Defects

R. Pfäffle, O. Blankenstein, S. Wüller, K. Heimann, G. Heimann

Department of Paediatrics, RWTH Aachen School of Medicine, Aachen, Germany

The pituitary gland develops in numerous discernable differentiation steps that finally lead to the formation of a complicated secretory organ. At least five different hormone-producing cell lines work in close proximity to each other under the control of separate hypothalamic regulatory centers.

During the embryological development of the anterior pituitary, various developmental factors are detectable with highly specific spatial and temporal expression patterns. These factors interact with each other either directly or by stimulation or repression of one another's expression; they form a complicated network of interdependent factors that guide the differentiation and proliferation of the hormone precursor cells and finally form the shape of the mature anterior pituitary lobe.

The primordium of the anterior pituitary is first visible at embryonic day (e) 8.5 in the mouse. The first steps of pituitary formation occur when a layer of ectoderm with single cell thickness invaginates cranially and contacts the primordium of the ventral hypothalamus. Thereafter, at e12.5 in the mouse, the anterior pituitary loses contact from the oral ectoderm and a period of intense cell proliferation and differentiation occurs. Finally five different cell lines have formed that are characterized by the specific hormone(s) that they produce: somatotrophs produce growth hormone (GH), lactotrophs prolactin (PRL), thyrotrophs TSH, gonadotrophs produce LH and FSH, and corticotrophs ACTH.

The different steps of pituitary cell formation are not as well studied in humans as they are in the mouse as the model organism. However, the principal steps of pituitary cell differentiation are presumably the same, because the pituitary differentiation factors involved are found highly conserved between species.

Frequency and Pathophysiology of Pituitary Hormone Deficiency

Pituitary hormone deficiency can affect one single cell line, as the somatotrophs in isolated growth hormone deficiency (IGHD), or multiple cell lines as in combined pituitary hormone deficiency (CPHD).

Estimates about the overall frequency of GHD, isolated or as a part of combined pituitary insufficiency, differ significantly and range from 1:4,000 to 1:10,000 [1, 2]. Trauma to the hypothalamo-pituitary system caused by tumor or infection can only account for the minority (10–20%) of cases [3, 4]. On the other hand, genetic causes of pituitary insufficiency appear to be rare. For the clinician the vast majority of the cases with pituitary insufficiency seem to occur sporadically, although in 5–30% of this group multiple occurrences in the family suggest a possible genetic background [5]. However, until now no systematic study has addressed this issue accurately, because many of the genetic factors involved in pituitary organogenesis have been unknown so far.

The Clinic of Combined Pituitary Hormone Deficiency

The differentiation of the pituitary somatolactotrophs into somatotrophs and lactotrophs is one of the last differentiation steps of hormone-producing cells in the pituitary. This is most likely the reason why in CPHD with a genetic origin, somatotrophs and lactotrophs are often affected together.

Thyrotrophs do also share part of the pituitary differentiation steps together with somatolactotrophs, which again explains why secondary hypothyroidism is a frequent additional finding in patients with GH and PRL deficiency.

Finally, although the regulation of the differentiated gonadotroph cells profoundly differs from that of somatotrophs, lactotrophs and thyrotrophs, these different cell lines share a common transcription factor (PROP1) during early stages of pituitary cell differentiation. Secondary hypogonadism therefore develops in patients with PROP1 gene defects, who are GH-deficient and also have secondary hypothyroidism.

In this respect, the examination of the PROP1 gene can be of significant clinical interest, because one of the most difficult questions for the treating physician remains whether the prepubertal child will develop secondary hypogonadism later on as part of hypopituitarism.

The severity of each of the pituitary hormone deficiencies however can vary quite considerably even among patients with similar genetic forms of CPHD. This chapter will try to give an account on which factors are relevant for the phenotypic expression of the disorders caused by POU1F1 and PROP1 gene defects.

Identification and Isolation of Pituitary Transcription Factors and Studies in Animal Models

Identification of Pit1 as a DNA-Binding Protein in the Mouse Pituitary

GH synthesis makes up the major proportion of protein synthesis within the anterior pituitary gland: approximately 3% of all pituitary-derived mRNA corresponds to the GH gene [6]. The pituitary gland has an average weight of approximately 500 mg and produces 1–3 mg of GH each day. Although GH synthesis can be detected in other tissues than the pituitary gland, this probably has no physiological relevance for the growth of the organism. The mechanisms responsible for the regulation of GH synthesis and secretion therefore have to be pituitary-specific and highly efficient at the same time.

GH gene transcription is regulated by numerous transcription factors, among which the pituitary transcription factor Pit1 plays an important role. Pit1 (pituitary transcription factor 1) has been given many different names, which include GHF1 (GH factor 1) and most recently Pit1. POU is an acronym of Pit1, Oct1 [7] and Unc86 [8], which are all related transcription factors from a variety of different organisms with similar DNA-binding domains.

Pit1 was identified in the mouse pituitary as a protein that binds to the promoter of the GH gene promoter [9, 10]. Due to this fact it initially was named growth hormone factor-1 (GHF1). Later however it was realized that Pit1 binds to AT-rich sequences that carry the common core sequence (TATNCAT). Pit1-binding sites have been isolated in the GH-gene promoter [11], the PRL gene promoter and enhancer [12] and the TSH gene promoter [13]. The binding sites for Pit1 within the GH gene promoter are highly conserved during evolution from fish to humans which underlines the functional importance of this pituitary transcription factor in mammalian development [14].

Pit1 expression therefore is not restricted to somatotrophs, but it is detectable in three distinct pituitary cell lines (somatotrophs, lactotrophs and thyrotrophs), which reflects its pluripotent role on cell differentiation during pituitary development. Pit1 is expressed in the mouse pituitary starting on e14.5 and the expression persists throughout adult life. Due to its high degree of conservation between the human and mouse homologues, this expression pattern would be expected to be similar during human embryogenesis, but of course on a different time scale.

Thyrotrophs seem to arise from two independent cell populations in mice [15]. The first cell population shows no Pit1 expression and seems to be only transiently present; these cells are detected on e12 at the rostral tip of the pituitary gland prior to the first detectable Pit1 expression on e14.5. The second population of thyrotrophs is found subsequently in the caudomedial portion of the anterior pituitary on e15.5 after Pit1 is expressed in this area [16]. These cells are Pit1-dependent.

Additionally, Pit1-binding sites have been identified in the GHRH-receptor (GHRHR) gene promoter [17] and the POU1F1 gene promoter itself [18]. The latter finding carries important implications as to an autostimulatory role of POU1F1 during pituitary organogenesis. However, *Pit1* gene transcription is detectable before the detection of Pit1 protein, which makes it unlikely that Pit1 induces its own developmental activation [19]. On the other hand, the disappearance of temporarily expressed Pit1 protein in the POU1F1 mutant dwarf Snell mice suggests that (intact) Pit1 protein is necessary for sustained *Pit1* gene expression [20].

Despite its role as a developmental factor and its early onset of expression during pituitary cell differentiation, POU1F1 is found in the somatotrophs, lactotrophs and thyrotrophs throughout life. This demonstrates its important role as a modulator of GH and PRL and most probably TSH synthesis and secretion.

Snell, Jackson and Ames Dwarf Mice

The various physiological and biochemical characteristics of Pit1 have been initially studied in the mouse model [15]. GH, PRL and TSH deficiency could be anticipated as the characteristic pituitary defect of Pit1-deficient individuals. It therefore seemed logical to investigate Snell, Jackson and Ames dwarf mice [21], that show pituitary hypoplasia with CPHD for these three hormones. Although the pituitaries of all three mouse strains are deficient for the Pit1 protein, mutations of the *Pit1* gene were found only in the Jackson (dw^J) and Snell dwarf mice [22]; no such changes could be detected in *Pit1* gene of Ames dwarf mice.

In the Jackson (dw^J) dwarf mouse an altered RFLP pattern of the *Pit1* gene suggested a gene alteration either by an inversion or insertion of a 4-kb fragment. The Snell (*dw*) *Pit1* gene showed a single nucleotide change C $\rightarrow$ G that converted a tryptophan located within the DNA-binding region to a cysteine (W261C).

Identification of Prop1 by Positional Cloning

Li et al. [22] realized that the Ames dwarf locus (*df*) was distinct to the one found in the Snell and Jackson dwarf mice; shortly thereafter, Buckwalter et al. [23] allocated this locus to mouse chromosome 11 which was different to the locus of the *Pit1* gene (mouse chromosome 16).

After some years, Sornson et al. [24] in a major effort exactly localized the *df* locus using advanced positional cloning techniques. They identified a factor with an expression during pituitary development limited to those cells that subsequently are found to express POU1F1. Sornson et al. therefore named the newly identified transcription factor the 'Prophet of Pit1' or Prop1.

Table 1. Transcription factors with a role in pituitary development (e = embryonic day)

Transcription factor	Time-scale of expression (in mouse pituitary)	Expression found in	Animal model for gene defect (mouse)	Known function
Prop1	e10–15	Somato-, lacto-, thyro- and gonadotrophs	Ames dwarf mouse	Proliferation of cells that later express POU1F1, differentiation of gonadotroph cells
Pit1 (POU1F1)	e13.5	Somato-, lacto- and thyrotrophs	Snell and Jackson dwarf mice	Differentiation and proliferation of somato-, lacto-, thyrotroph cells

Prop1 is first detected at e10 in the mouse pituitary and is mainly detectable over the dorsal portion of the gland. A maximum of Prop1 expression is detected by e12 in the caudomedial region, a region where Pit1 is expressed later on [25]. Analysis of double mutants revealed that the growth-retarded phenotype of the Ames mouse prevails over the Snell and Jackson mouse phentotypes.

The Ames dwarf mouse has a homozygous mutation which involves a serine to proline substitution at amino acid 83 (S83P) in the first α helix of the homeodomain. These animals have a deficiency for GH, PRL and TSH, however show normal initial stages of pituitary development. At e13 in the mouse, pituitary cells in the caudomedial region fail to express Pit1 and thereby Pit1-dependent pituitary lineages do not differentiate and proliferate. The size of the pituitary gland is visibly reduced by e14.5, and the size of the adult pituitary is reduced by approximately two thirds.

The adult Ames dwarf mouse shows severe GH deficiency with less than 1% of the normal somatotrophs, and scarcity of lactotrophs and thyrotrophs. There is also reduced expression of gonadotropins (table 1).

Structure and Function of POU1F1 and PROP1

The human POU1F1 (PIT1) gene is located on chromosome 3p11 and consists of six exons spanning a region of 17 kb of DNA sequence. The gene encodes a protein of 291 amino acids with a molecular mass of 33 kDa. The protein has three functional domains which include a transactivation domain and two DNA-binding domains – a POU-specific domain and a POU homeodomain. The POU homeodomain has a nearly identical structure to that seen in many of the homeobox genes, i.e. a highly conserved region of 180 nucleotides encoding a 60-amino-acid DNA-binding domain within the protein [8].

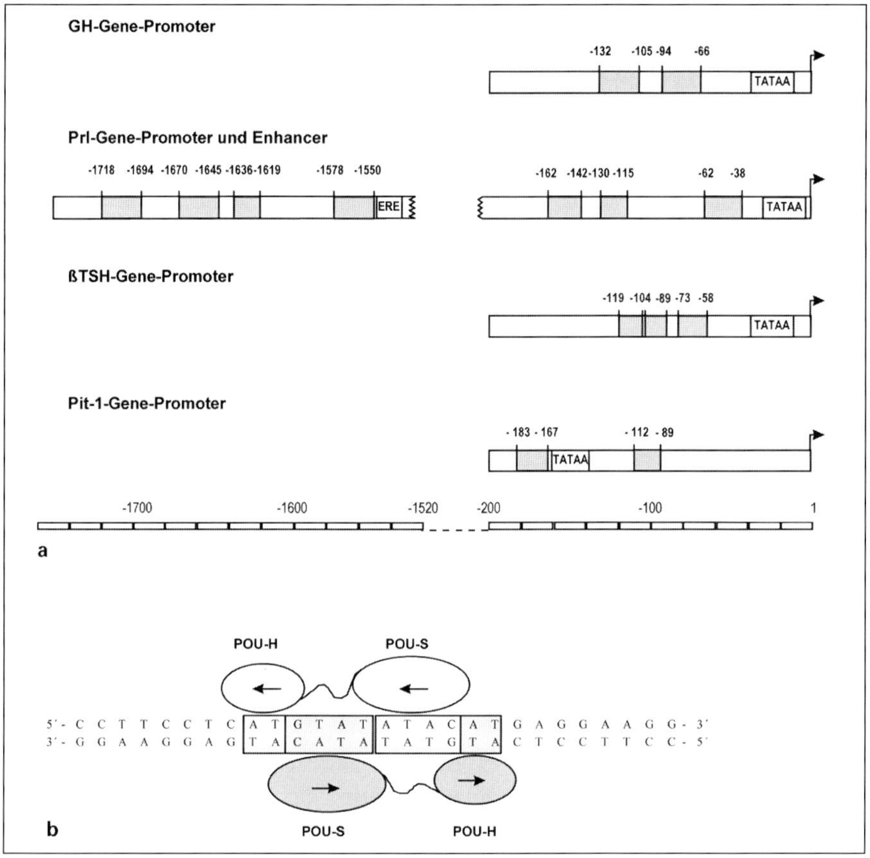

Fig. 1. a Schematic view of the binding sites within the known target genes of POU1F1. The arrow indicates the transcriptional start site of the target gene. The TATA box and possible estrogen response elements (ERE) have been marked. The consensus sequence for POU1F1 is TATNCAT, that within the TSH-β gene however is AATNCAT. ***b*** POU1F1 binds as a homodimer to the target sequence contacting opposite sides of the DNA double helix. Because the orientation of the POU1F1 molecule is antiparallel, the homeo subdomain (POU-H) lies next to the POU specific of the opposite POU1F1 molecule. Mutations within the DNA-binding domain of POU1F1 can interfere both with the DNA contact and the contacts made between the two proteins.

The POU homeodomain specifically recognizes and binds to A-T-rich regions in DNA, while the POU-specific domain enhances this by binding cooperatively at the consensus sequence TATNCAT (where N is an unspecified nucleotide) (fig. 1b) [26–28]. The promoter regions of the genes encoding GH and CSH have two such POU1F1 consensus binding sites. Three POU1F1-binding sites

Table 2. Characteristics of the human pituitary transcription factors POU1F1 and PROP1

	Gene (human)			Protein (human)			
	gene locus	length of gene kb	exons n	length (amino acids)	molecular weight kDa	DNA-binding domain	binding site (consensus sequence)
PROP1	5q	3	3	226		Paired like homeo	TATNCAT
POU1F1 (PIT1)	3p11	17	6	291	33	POU-homeo	TAATTGAATTA[a]

[a] Proposed sequence of PROP1-binding sites as deduced from structure of PROP1 homeodomain by 'random site selection' [58]; physiological target genes of PROP1 are unknown so far.

have also been reported in the PRL promoter. Interestingly, binding sites have also been demonstrated in the promoter region of POU1F1 (fig. 1a).

PROP1 is a 'paired'-like homeobox gene, which has been mapped to chromosome 5q in the human genome [29]. The term Prop1 is an acronym of Prophet of PIT1. This nomenclature was used because the Prop1 protein product precedes and promotes the expression of Pit1.

The *PROP1* gene consists of three exons, which span a region of approximately 3 kb. This human gene encodes for a protein of 226 amino acids. The transcription factor has one central DNA-binding domain with three α helical regions, which resembles the 'paired' domains that are characteristic of transcription factors found in *Drosophila*. In the fruitfly, these factors participate in determining cell fate of various cell lines and organs during embryonic development. In addition, a transcriptional activation domain can be delineated by mutational analyses, which is involved in the autoregulatory control of Prop1 expression. Similar to other 'paired'-like nuclear proteins, Prop1 'recognizes' and binds to palindromic DNA sequences that usually can be found within the promoters of other target genes. In vivo the protein binds to these sequences as a dimer [30] (table 2).

Clinical Pictures Caused by the Gene Defects

Mutations of POU1F1

The first human POU1F1 mutation was reported in 1992 in a child with combined GH and PRL deficiency [31]. The patient showed also profound hypothyroidism, which suggested a defect of TSH production. The patient displayed homozygosity for a nonsense mutation at codon 172, which results in the production of a severely truncated POU1F1 protein. The mutant protein lacks both segments of the DNA-binding domain and therefore will show no binding to GH, PRL and TSH promoters. Therefore, this mutant protein is unlikely to interfere with target gene transcription.

A rather uniform hormonal phenotype is present in patients with POU1F1 mutations: CPHD with lack of GH, PRL and TSH secretion can be observed in all patients after 1–2 years of life [32]. GH and PRL serum levels are found below detection limits of the common RIA assays used in clinical diagnosis. TSH serum levels are usually also low, however can be found within the lower normal range in patients with POU1F1 defects during their first years of life.

Whereas GH deficiency seems unequivocal in all patients with POU1F1 deficiency, hypothyroidism therefore can be more or less severe. This might obscure the initial diagnosis of hypothyroidism in some of these patients [33]. On the other hand, hypothyroidism may be the leading clinical symptom in patients with POU1F1 defect right after birth [34, 35], despite the fact that in others it occurs only after initiation of GH substitution therapy. This however might only reflect the fact that GH deficiency masks hypothyroidism. So far, no explanation has been found for the variability in the clinical severity of hypothyroidism. It cannot be attributed to the exact site of the mutation as identical mutations have been observed to produce variable degrees of hypothyroidism at the time of diagnosis of the disorder.

From birth on, severe PRL deficiency most likely is present in all patients with *POU1F1* gene mutations, however lack of symptoms usually makes this an incidental finding during testing procedures for pituitary insufficiency. Women with POU1F1 defects have had uneventful pregnancies if substituted, however were unable to breast-feed their children [32, 35].

CPHD due to *POU1F1* gene mutations has been shown to be inherited in a recessive mode [31, 33]. In other pedigrees the disorder follows an autosomal dominant mode of inheritance [36, 37].

POU1F1 mutations that cause structural changes within the POU domain of POU1F1 are inherited in an autosomal recessive way. This is probably due to the fact that most mutations which affect the structure of the helix-turn-helix motif impair the sequence-specific binding of the protein to the DNA target sites.

POU1F1 mutations outside the DNA-binding domain follow an autosomal dominant way of inheritance as some of them have been shown to not impair or even to enhance the binding of POU1F1 to the DNA target sequences. This would be an explanation for the dominant negative effect of these mutations [36].

Mutations within the *POU1F1* gene are scattered throughout the portion of the gene that codes for the DNA-binding domain of the POU1F1 protein. Over the years however, one 'hotspot' for mutations within the *POU1F1* gene has emerged. The R271W mutation with an autosomal dominant type of inheritance

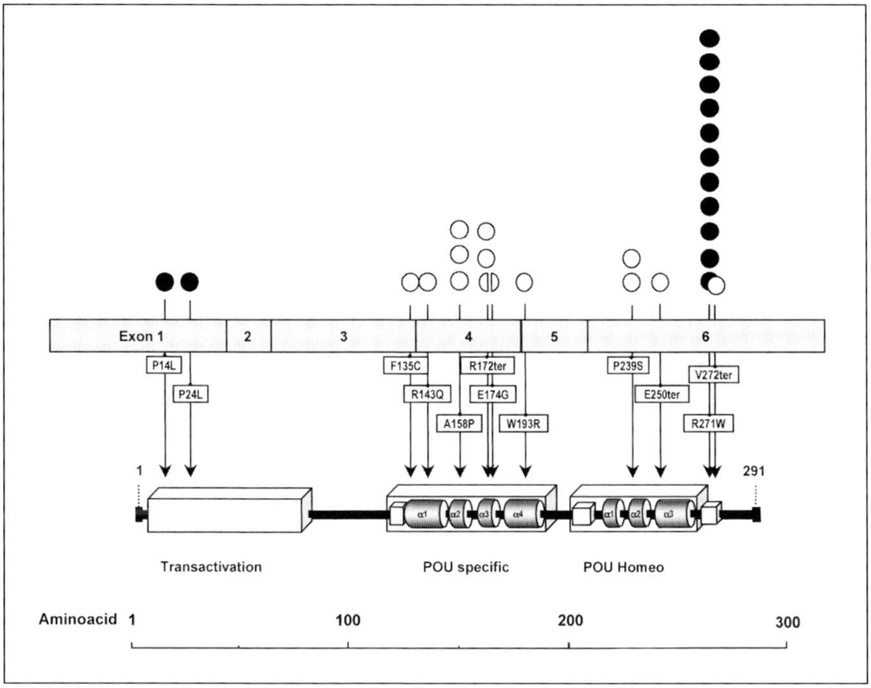

Fig. 2. Mutations found within the human *POU1F1* gene. The arrows show the location of the mutation in respect to the 6 coding exons and the relative position within the POU1F1 protein. Closed circles stand for the location of mutations that show an autosomal dominant type of inheritance, open circles indicate the position of mutations with a autosomal recessive inheritance. Half circles indicate the position of each mutation found in patients with compound heterozygosity for two recessive mutations.

has been observed sporadically and in pedigrees from various ethnic origins [38–44]. This may be due to the DNA sequence at this specific site of the gene, a so-called CpG dinucleotide which is prone to spontaneous mutations. In keeping with its dominant negative effect, this mutation is responsible for the clinical manifestation if only one allele is affected. The variable *POU1F1* gene defects described so far are depicted in figure 2.

Mutations in the PROP1 Gene

After the description of the mouse Prop1 gene and the phenotype of the Ames dwarf mice [24], patients with a CPHD of GH, PRL and TSH were screened for PROP1 mutations and no defects were identified. Patients with a wider spectrum of pituitary hormone deficiencies however showed *PROP1* gene mutations [30]. This indicated the phenotypic difference between the dwarf Ames mouse and patients with PROP1 mutations, although a slightly insufficient gonadotropin secretion has also been observed in Ames mice [24].

PROP1 gene mutations induce CPHD that includes LH and FSH in addition to the GH, PRL and TSH deficiency observed in patients with POU1F1 gene defects [30].

A severe deficiency for GH and PRL can be detected in patients with PROP1 mutations, nevertheless in contrast to patients with POU1F1 mutations, the conventional RIAs can detect remnant levels of hGH and PRL in patients with PROP1 mutations.

Similar to some patients with POU1F1 mutations, TSH deficiency may not be present in all persons with *PROP1* gene defect right after birth. When measured during early infancy, TSH serum levels are still detectable, sometimes even in the lower normal range. However, if measured after TRH stimulation the TSH rise is delayed, resembling tertiary or hypothalamic hypothyroidism. A similar pattern can be observed for the function of the gonadotroph cell line. Secondary hypogonadism is a preeminent symptom of all adult patients with *PROP1* gene defects, however almost always hypogonadism is not encountered before the time of puberty in patients with PROP1 mutations.

The variability of this facet of the CPHD is evidenced by the fact that some patients with PROP1 mutations showed initial stages of pubertal development before secondary hypogonadism manifested [45]. Even though patients with POU1F1 mutations also show different degrees of hypothyroidism at diagnosis, there is greater variability in the clinical manifestation of PROP1 mutations during the first years of life. Considering the fact that such differences can be observed even between patients with identical PROP1 mutations, the mechanisms for this phenomenon are not clear yet.

One observation made in the dwarf mouse model could reflect the lack of completeness of pituitary hormone deficiency observed in human PROP1 mutations. The pituitaries of Ames dwarf mice do not show complete lack of Pit1 expression as observed in the Snell and Jackson mice [24]. In contrast to the Snell dwarf mouse where no somatotrophs can be detected in the anterior pituitary, between 0.001% [46] and 1% [47] of the number of somatotrophs found in wild-type mice can be detected in the Ames dwarf mice. Some somatotrophs, lactotrophs and thyrotrophs might escape the influence of the mutant Prop1 and progress to further proliferation and differentiation.

Due to its temporary expression, the differentiating effects of Prop1 are exerted only during a limited time of pituitary development. Pit1 in contrast exerts its transactivating properties on target gene transcription also in the adult individual. This fact could add to the profoundness of the Pit1 deficiency phenotype.

Patients with PROP1 mutations have also been reported to display some minor degree of hypocortisolism that requires steroid substitution. However, *PROP1* is not reported to be expressed in corticotrophs. This finding is mainly made in adult patients with *PROP1* gene defects and it might reflect a decreasing corticotroph function in the context of an expanding pituitary insufficiency [48].

PROP1 gene mutations are inherited in an autosomal recessive way, so far no mutations with a dominant negative effect have been observed.

Two mutations within the *PROP1* gene have been observed with increased frequency. These mutations are both microdeletions within exon 2 of the *PROP1* gene. The 2-bp ΔGA301/2 deletion [30] and the ΔGA149/50 or ΔA150 frameshift mutations introduce early stop codons at positions 109 and 164 respectively [49]. The predicted PROP1 protein molecules resulting from these two mutations do not contain sequences corresponding to the highly conserved DNA-binding domain. A nonsense point mutation, R99X, found within exon 2 has been observed only once [50]. All missense mutations like the F117I mutation, the R120C mutation [30], the F88S [51] and the R73C PROP1 mutation [29], found within the *PROP1* gene so far, affect the DNA-binding domain of the protein (fig. 3).

In patients with PROP1 mutations a combination of GH, PRL, TSH and gonadotropin deficiency is detectable, the age of the patients at the time of diagnosis however varies [45] (table 3).

Pituitary Morphology in POU1F1- and PROP1-Deficient Patients

The adult pituitary gland differs considerably in size and weight (400–900 mg). Although MR tomography provides an excellent picture of hypothalamo-pituitary anatomy, it is not possible to judge from pituitary size its functional capacity. Images of pituitaries of patients with POU1F1 or PROP1 mutations show small (approximately one third of normal) anterior pituitary glands compared with those of age matched healthy individuals [52].

No significant size difference can be seen between the size of the pituitary glands from patients with POU1F1 mutations compared with those of patients with PROP1 mutations [53].

MR images of the pituitary are essential to exclude space-occupying lesions within the hypothalamo-pituitary axis. Anatomical defects like midline lesions or pituitary ectopia are not seen in patients with *POU1F1* and *PROP1* gene defects [54], such findings however are associated with genetic defects of factors that influence earlier stages of pituitary development like HesX-1 [55].

In approximately one third of the patients with *PROP1* gene defects, pituitary tumors have been observed [56, 57]. These intra- or suprasellar space-occupying lesions are usually found incidentally during diagnostic workup. They have never been reported to result in secondary sequelae like visual loss or head-

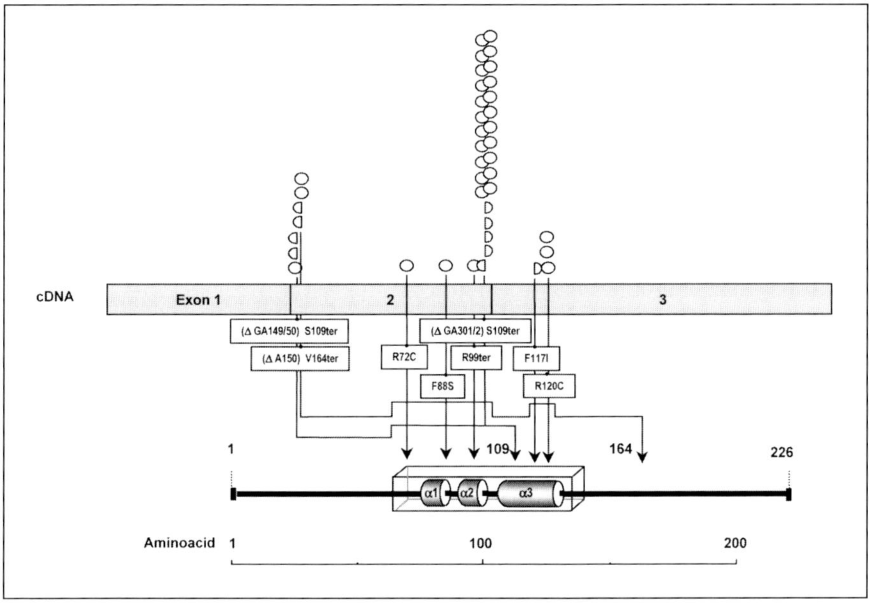

Fig. 3. Location of PROP1 mutations in the human PROP1 gene. Two mutations have been identified that occur frequently in patients with CPHD. Both 'hotspot' mutations (A150 and GA301/2) induce frameshifts within the open reading frame (symbolized by the horizontal part of the line) and introduce early translational stop codons (positions 109 and 164 respectively). All mutations found within the human PROP1 gene so far are located in exons 2 and 3, and affect the DNA-binding domain of the PROP1 protein (for further explanations see text). The mutations observed so far have shown an autosomal recessive type of inheritance.

Table 3. Clinical phenotype of POU1F1 and PROP1 mutations with suggested investigations for clinical diagnosis

Transcription factor	Hormonal phenotype (deficiencies for)	Inheritance	Pituitary morphology
POU1F1	GH, PRL, TSH (GH and PRL deficiency is profound, dependent on age TSH levels can vary → TRH test may be necessary)	Dominant or recessive	Reduced size
PROP1	GH, PRL, TSH, LH/FSH (GH and PRL deficiency is profound, TSH and gonadotropin levels can vary → TRH and LHRH tests are recommended; in some instances a slight degree of ACTH deficiency can develop over years → CRF testing optional)	Recessive	Reduced size Sometimes intra- or suprasellar enlargement

aches. Their presence therefore is more irritating for the attending physician than for the patient.

In the follow-up of some of the patients it became evident that these tumors regress after years of persistent enlargement within relatively short periods of time. Surgery therefore cannot be recommended. This phenomenon is not observed in the Prop1 mutant Ames mouse, which again indicates the substantial differences that exist between the animal model and its human equivalents. So far, one can only speculate about the origin of these cell masses. As surgery is hardly indicated in face of the spontaneous course of these lesions, information on the histology of these pituitary masses will remain scarce.

One hypothesis is that the defect PROP1 might fail to repress some early pituitary transcription factors. This prolonged expression may give rise to such pituitary enlargement. A prolonged expression of Lhx3 has been observed for example in the Ames mouse [24].

Frequency of POU1F1 and PROP1 Gene Defects

Genetic analyses cannot explain all forms of CPHD, in our own experience with more than 75 patients showing CPHD of various forms, approximately half of all patients with a deficiency for GH, TSH and PRL have detectable mutations within their *POU1F1* gene. In contrast, two thirds of all patients with a deficiency of GH, TSH, PRL and gonadotropins have mutations within their *PROP1* gene. The efficacy of this genetic testing thereby greatly relies on the accuracy of the pituitary stimulation tests performed by the clinician.

The experience with transcription factors of early pituitary development (like HESX1 and LHX3) is still too limited to specify a definite hormonal phenotype or to get a good idea about the frequency of such mutations. Our preliminary results of the genetic screening in the cohort of 75 otherwise not preselected patients with various forms of CPHD suggest a rare incidence of mutations within these two genes, accounting for presumably less than 2% of patients with CPHD.

Although the genetic analysis of factors involved in pituitary differentiation is still costly and time-consuming, the clinical information obtained with the detection of such mutations is mounting. This creates an increasingly detailed picture of the different clinical courses that these genetic defects can take in the CPHD patients. Finally, this knowledge will facilitate an appropriate and cost-effective clinical management and thereby justify such expenditure.

References

1 Lacey KA, Parkin JM: Causes of short stature. A community study of children in Newcastle upon Tyne. Lancet 1974;i:42–45.

2 Vimpani GV, Vimpani AF, Lidgard GP, Cameron EH, Farquhar JW: Prevalence of severe growth hormone deficiency. Br Med J 1977;ii:427–430.

3 Phillips JA: Inherited defects in growth hormone synthesis and action; in Scriver CR, Beaudet AL, Sly WS, Valle D (eds): The Metabolic Basis of Inherited Disease. New York, McGraw-Hill, 1995, pp 3023–3044.

4 Wilton P, Wallstrom A: An overview of the diagnoses in the Kabi Pharmacia International Growth Study. Acta Paediatr Scand Suppl 1991;379:93–98.

5 Procter AM, Phillips JA 3rd, Cooper DN: The molecular genetics of growth hormone deficiency. Hum Genet 1998;103:255–272.

6 Chen EY, Liao YC, Smith DH, Barrera-Saldana HA, Gelinas RE, Seeburg PH: The human growth hormone locus: Nucleotide sequence, biology, and evolution. Genomics 1989;4:479–497.

7 Sturm RA, Das G, Herr W: The ubiquitous octamer-binding protein Oct-1 contains a POU domain with a homeo box subdomain. Genes Dev 1988;2:1582–1599.

8 Herr W, Sturm RA, Clerc RG, Corcoran LM, Baltimore D, Sharp PA, Ingraham HA, Rosenfeld MG, Finney M, Ruvkun G et al: The POU domain: A large conserved region in the mammalian pit-1, oct-1, oct-2, and *Caenorhabditis elegans* unc-86 gene products. Genes Dev 1988;2:1513–1516.

9 Ingraham HA, Chen RP, Mangalam HJ, Elsholtz HP, Flynn SE, Lin CR, Simmons DM, Swanson L, Rosenfeld MG: A tissue-specific transcription factor containing a homeodomain specifies a pituitary phenotype. Cell 1988;55:519–529.

10 Bodner M, Castrillo JL, Theill LE, Deerinck T, Ellisman M, Karin M: The pituitary-specific transcription factor GHF-1 is a homeobox-containing protein. Cell 1988;55:505–518.

11 Mangalam HJ, Albert VR, Ingraham HA, Kapiloff M, Wilson L, Nelson C, Elsholtz H, Rosenfeld MG: A pituitary POU domain protein, Pit-1, activates both growth hormone and prolactin promoters transcriptionally. Genes Dev 1989;3:946–958.

12 Ingraham HA, Albert VR, Chen RP, Crenshaw EB 3rd, Elsholtz HP, He X, Kapiloff MS, Mangalam HJ, Swanson LW, Treacy MN, et al: A family of POU-domain and Pit-1 tissue-specific transcription factors in pituitary and neuroendocrine development. Annu Rev Physiol 1990;52:773–791.

13 Steinfelder HJ, Hauser P, Nakayama Y, Radovick S, McClaskey JH, Taylor T, Weintraub BD, Wondisford FE: Thyrotropin-releasing hormone regulation of human TSHB expression: Role of a pituitary-specific transcription factor (Pit-1/GHF-1) and potential interaction with a thyroid hormone-inhibitory element. Proc Natl Acad Sci USA 1991;88:3130–3134.

14 Argenton F, Vianello S, Bernardini S, Jacquemin P, Martial J, Belayew A, Colombo L, Bortolussi M: The transcriptional regulation of the growth hormone gene is conserved in vertebrate evolution. Biochem Biophys Res Commun 1993;192:1360–1366.

15 Simmons DM, Voss JW, Ingraham HA, Holloway JM, Broide RS, Rosenfeld MG, Swanson LW: Pituitary cell phenotypes involve cell-specific Pit-1 mRNA translation and synergistic interactions with other classes of transcription factors. Genes Dev 1990;4:695–711.

16 Treier M, Gleiberman AS, O'Connell SM, Szeto DP, McMahon JA, McMahon AP, Rosenfeld MG: Multistep signaling requirements for pituitary organogenesis in vivo. Genes Dev 1998;12:1691–1704.

17 Petersenn S, Rasch AC, Heyens M, Schulte HM: Structure and regulation of the human growth hormone-releasing hormone receptor gene. Mol Endocrinol 1998;12:233–247.

18 Chen RP, Ingraham HA, Treacy MN, Albert VR, Wilson L, Rosenfeld MG: Autoregulation of pit-1 gene expression mediated by two cis-active promoter elements. Nature 1990;346:583–586.

19 Lew D, Brady H, Klausing K, Yaginuma K, Theill LE, Stauber C, Karin M, Mellon PL: GHF-1-promoter-targeted immortalization of a somatotropic progenitor cell results in dwarfism in transgenic mice. Genes Dev 1993;7:683–693.

20 Rhodes SJ, Chen R, DiMattia GE, Scully KM, Kalla KA, Lin SC, Yu VC, Rosenfeld MG: A tissue-specific enhancer confers Pit-1-dependent morphogen inducibility and autoregulation on the pit-1 gene. Genes Dev 1993;7:913–932.

21 Duquesnoy RJ: The pituitary dwarf mouse: A model for study of endocrine immunodeficiency disease. Birth Defects Orig Artic Ser 1975;11:536–543.
22 Li S, Crenshaw EB, Rawson EJ, Simmons DM, Swanson LW, Rosenfeld MG: Dwarf locus mutants lacking three pituitary cell types result from mutations in the POU-domain gene pit-1. Nature 1990; 347:528–533.
23 Buckwalter MS, Katz RW, Camper SA: Localization of the panhypopituitary dwarf mutation (df) on mouse chromosome 11 in an intersubspecific backcross. Genomics 1991;10:515–526.
24 Sornson MW, Wu W, Dasen JS, Flynn SE, Norman DJ, SM OC, Gukovsky I, Carriere C, Ryan AK, Miller AP, Zuo L, Gleiberman AS, Andersen B, Beamer WG, Rosenfeld MG: Pituitary lineage determination by the Prophet of Pit-1 homeodomain factor defective in Ames dwarfism. Nature 1996;384:327–333.
25 Dutour A: A new step understood in the cascade of tissue-specific regulators orchestrating pituitary lineage determination: The Prophet of Pit-1 (Prop-1). Eur J Endocrinol 1997;137:616–617.
26 Theill LE, Castrillo JL, Wu D, Karin M: Dissection of functional domains of the pituitary-specific transcription factor GHF-1. Nature 1989;342:945–948.
27 Verrijzer CP, Alkema MJ, Van Weperen WW, Van Leeuwen HC, Strating MJ, Van der Vliet PC: The DNA binding specificity of the bipartite POU domain and its subdomains. EMBO J 1992;11: 4993–5003.
28 Assa-Munt N, Mortishire-Smith RJ, Aurora R, Herr W, Wright PE: The solution structure of the Oct-1 POU-specific domain reveals a striking similarity to the bacteriophage lambda repressor DNA-binding domain. Cell 1993;73:193–205.
29 Duquesnoy P, Roy A, Dastot F, Ghali I, Teinturier C, Netchine I, Cacheux V, Hafez M, Salah N, Chaussain JL, Goossens M, Bougneres P, Amselem S: Human Prop-1: Cloning, mapping, genomic structure. Mutations in familial combined pituitary hormone deficiency. FEBS Lett 1998;437:216–220.
30 Wu W, Cogan JD, Pfäffle RW, Dasen JS, Frisch H, O'Connell SM, Flynn SE, Brown MR, Mullis PE, Parks JS, Phillips JA, Rosenfeld MG: Mutations in PROP1 cause familial combined pituitary hormone deficiency. Nat Genet 1998;18:147–149.
31 Tatsumi K, Miyai K, Notomi T, Kaibe K, Amino N, Mizuno Y, Kohno H: Cretinism with combined hormone deficiency caused by a mutation in the POU1F1 gene. Nat Genet 1992;1:56–58.
32 Pfäffle RW, Kim C, Otten B, Wit JM, Eiholzer U, Heimann G, Parks J: Pit-1: Clinical aspects. Horm Res 1996;45:25–28.
33 Pfäffle RW, DiMattia GE, Parks JS, Brown MR, Wit JM, Jansen M, Van der Nat H, Van den Brande JL, Rosenfeld MG, Ingraham HA: Mutation of the POU-specific domain of Pit-1 and hypopituitarism without pituitary hypoplasia. Science 1992;257:1118–1121.
34 Pfäffle RW, Martinez R, Kim C, Frisch H, Lebl J, Otten B, Heimann G: GH and TSH deficiency. Exp Clin Endocrinol Diabetes 1997;105(suppl 4):1–5.
35 De Zegher F, Pernasetti F, Vanhole C, Devlieger H, Van den Berghe G, Martial JA: The prenatal role of thyroid hormone evidenced by fetomaternal Pit-1 deficiency. J Clin Endocrinol Metab 1995; 80:3127–3130.
36 Radovick S, Nations M, Du Y, Berg LA, Weintraub BD, Wondisford FE: A mutation in the POU-homeodomain of Pit-1 responsible for combined pituitary hormone deficiency. Science 1992;257: 1115–1118.
37 Ohta K, Nobukuni Y, Mitsubuchi H, Fujimoto S, Matsuo N, Inagaki H, Endo F, Matsuda I: Mutations in the Pit-1 gene in children with combined pituitary hormone deficiency. Biochem Biophys Res Commun 1992;189:851–855.
38 Aarskog D, Eiken HG, Bjerknes R, Myking OL: Pituitary dwarfism in the R271W Pit-1 gene mutation. Eur J Pediatr 1997;156:829–834.
39 Arnhold IJ, Nery M, Brown MR, Voss TC, VanderHeyden TC, Adess ME, Hurley DL, Wajchenberg BL, Parks JS: Clinical and molecular characterization of a Brazilian patient with Pit-1 deficiency. J Pediatr Endocrinol Metab 1998;11:623–630.
40 Cohen LE, Wondisford FE, Salvatoni A, Maghnie M, Brucker-Davis F, Weintraub BD, Radovick S: A 'hot spot' in the Pit-1 gene responsible for combined pituitary hormone deficiency: Clinical and molecular correlates. J Clin Endocrinol Metab 1995;80:679–684.

41 Rodrigues Martineli AM, Braga M, De Lacerda L, Raskin S, Graf H: Description of a Brazilian patient bearing the R271W Pit-1 gene mutation. Thyroid 1998;8:299–304.
42 Ward I, Chavez M, Huot C, Lecocq P, Martial JA, Van Vliet G: Severe combined pituitary hormone deficiency due to a Pit-1 mutation in a French-Canadian boy (abstract). Horm Res 1997;48: 1–201.
43 Radovick S, Cohen LE, Wondisford FE: The molecular basis of hypopituitarism. Horm Res 1998; 49:30–36.
44 Pfäffle RW, Parks JS, Brown MR, Heimann G: Pit-1 and pituitary function. J Pediatr Endocrinol 1993;6:229–233.
45 Fluck C, Deladoey J, Rutishauser K, Eble A, Marti U, Wu W, Mullis PE: Phenotypic variability in familial combined pituitary hormone deficiency caused by a PROP1 gene mutation resulting in the substitution of Arg → Cys at codon 120 (R120C). J Clin Endocrinol Metab 1998;83:3727–3734.
46 Andersen B, Pearse RV, Jenne K, Sornson M, Lin SC, Bartke A, Rosenfeld MG: The Ames dwarf gene is required for Pit-1 gene activation. Dev Biol 1995;172:495–503.
47 Gage PJ, Brinkmeier ML, Scarlett LM, Knapp LT, Camper SA, Mahon KA: The Ames dwarf gene, df, is required early in pituitary ontogeny for the extinction of Rpx transcription and initiation of lineage-specific cell proliferation. Mol Endocrinol 1996;10:1570–1581.
48 Asteria C, Oliveira JH, Abucham J, Beck-Peccoz P: Central hypocortisolism as part of combined pituitary hormone deficiency due to mutations of PROP-1 gene. Eur J Endocrinol 2000;143:347–352.
49 Fofanova O, Takamura N, Kinoshita E, Parks JS, Brown MR, Peterkova VA, Evgrafov OV, Goncharov NP, Bulatov AA, Dedov II, Yamashita S: Compound heterozygous deletion of the PROP-1 gene in children with combined pituitary hormone deficiency. J Clin Endocrinol Metab 1998;83: 2601–2604.
50 Parks JS, Brown MR, Hurley DL, Phelps CJ, Wajnrajch MP: Heritable disorders of pituitary development. J Clin Endocrinol Metab 1999;84:4362–4370.
51 Osorio MG, Kopp P, Marui S, Latronico AC, Mendonca BB, Arnhold IJ: Combined pituitary hormone deficiency caused by a novel mutation of a highly conserved residue (F88S) in the homeodomain of PROP-1. J Clin Endocrinol Metab 2000;85:2779–2785.
52 Fofanova O, Takamura N, Kinoshita E, Vorontsov A, Vladimirova V, Dedov II, Peterkova V, Yamashita S: MR imaging of the pituitary gland in children and young adults with congenital combined pituitary hormone deficiency associated with PROP1 mutations. AJR Am J Roentgenol 2000;174:555–559.
53 Pfäffle RW, Blankenstein O, Wuller S, Kentrup H: Combined pituitary hormone deficiency: Role of Pit-1 and Prop-1. Acta Paediatr Suppl 1999;88:33–41.
54 Sloop KW, Walvoord EC, Showalter AD, Pescovitz OH, Rhodes SJ: Molecular analysis of LHX3 and PROP-1 in pituitary hormone deficiency patients with posterior pituitary ectopia. J Clin Endocrinol Metab 2000;85:2701–2708.
55 Dattani MT, Martinez-Barbera JP, Thomas PQ, Brickman JM, Gupta R, Martensson IL, Toresson H, Fox M, Wales JK, Hindmarsh PC, Krauss S, Beddington RS, Robinson IC: Mutations in the homeobox gene HESX1/Hesx1 associated with septo-optic dysplasia in human and mouse. Nat Genet 1998;19:125–133.
56 Mendonca BB, Osorio MG, Latronico AC, Estefan V, Lo LS, Arnhold IJ: Longitudinal hormonal and pituitary imaging changes in two females with combined pituitary hormone deficiency due to deletion of A301,G302 in the PROP1 gene. J Clin Endocrinol Metab 1999;84:942–945.
57 Rosenbloom AL, Almonte AS, Brown MR, Fisher DA, Baumbach L, Parks JS: Clinical and biochemical phenotype of familial anterior hypopituitarism from mutation of the PROP1 gene. J Clin Endocrinol Metab 1999;84:50–57.
58 Wilson D, Sheng G, Lecuit T, Dostatni N, Desplan C: Cooperative dimerization of paired class homeo domains on DNA. Genes Dev 1993;7:2120–2134.

R.W. Pfäffle, Department of Paediatrics, RWTH Aachen School of Medicine
D–52074 Aachen (Germany)
Tel. +49 241 80 89892, Fax +49 241 8888 592, E-Mail rpfaeffle@post.klinikum.rwth-aachen.de

Rappaport R, Amselem S (eds): Hypothalamic-Pituitary Development. Genetic and Clinical Aspects.
Endocr Dev. Basel, Karger, 2001, vol 4, pp 77–93

Septo-Optic Dysplasia and Related Malformations

M.T. Dattani

London Centre for Paediatric Endocrinology, Institute of Child Health, London, UK

Development of the Forebrain and Anterior Pituitary Gland

Animal studies have recently shed a considerable amount of light on the normal development of the forebrain and pituitary gland. Evidence has now emerged for anterior patterning in the extraembryonic endoderm (anterior visceral endoderm – AVE) of the early mammalian conceptus before there is any sign of primitive streak formation, i.e. before gastrulation [1]. This AVE has been shown to be functional in providing the embryo with anterior character and is therefore vital for normal forebrain development. Expression of the genes *Otx2*, *Lim1*, *goosecoid*, *cerberus-related 1* and *Hex* is restricted to a medial strip of the AVE underlying approximately the anterior third of the epiblast at least 12 h before the primitive streak has formed. As the streak forms, the most anterior extreme of the AVE in the region where the heart will develop starts to express *Mrg1* and slightly more posteriorly, where it overlies epiblast fated to give rise to oral ectoderm and forebrain, *Hesx1* is expressed. Hence, as gastrulation starts, the AVE is already a genetically patterned tissue. Mutations in a number of genes which are first expressed in the AVE and only subsequently in epiblast derivatives affect anterior development in the mouse, with variable defects of forebrain formation. These include the paired-like homeobox gene *Hesx1, Lim1* and *Otx2.*

In man, the first event in the development of the nervous system is neural induction or neurulation, whereby the neural plate, the embryonic precursor of the brain, forms in the third week of gestation. This is the period known as gastrulation, whereby the primitive streak and the mesoderm of the embryo are formed. The neural plate forms as a result of the cellular interaction between the ectoderm and the mesoderm of the gastrula-stage embryo and is a planar sheet of pseudostratified neuroepithelium produced during gastrulation. The neural plate becomes regionally patterned along each of its axes and is converted into the neural tube which then slowly differentiates into the central nervous system consisting of the brain and spinal cord. Patterns of differential gene expression are evident at the anterior neural plate stage, and these foreshadow the later development of morphological and functional subdivisions of the neural tube. In the fourth week of gestation, a series of ring-like constrictions mark the approximate boundaries between the primordia of the major brain regions: the forebrain or prosencephalon, midbrain or mesencephalon and hindbrain or rhombencephalon. As the rostral neuropore closes at about 25 days of expression, two lateral outgrowths called the optic vesicles appear, one on each side of the forebrain. The optic vesicles are the primordia of the retinae and optic nerves. A second pair of diverticula arise more dorsally and rostrally; these are the cerebral or telencephalic vesicles which are the primordia of the cerebral hemispheres, and their cavities form the lateral ventricles. During the fifth week, the forebrain partly divides into two regions, the telencephalon and the diencephalon. The diencephalon will give rise to the thalamus and hypothalamus.

Fate map studies in a number of species such as the chick, the *Xenopus*, axolotl and zebrafish have shown that the prospective telencephalon occupies the rostral end of the neural plate, partly overlapping the anterior neural ridge (ANR). Proceeding laterally along the ANR, a sequence of sites are found which are fated to form the olfactory bulb and the lateral, dorsal and medial cortex. The neural plate medial to the telencephalon contains sites destined to form the preoptic, optic and hypothalamic areas. Of note is the finding that the anterior pituitary or adenohypophysis maps onto the ANR. Additionally, in the *Xenopus*, the mid-anterior ridge produces a triangular wedge of tissue extending at the midline from the chiasma ridge to the lamina terminalis and laterally into the optic stalks [for review, see 2]. Anterior to the notochord is an axial mesendodermal structure named the prechordal plate. Several lines of molecular and genetic evidence now suggest that medial and ventral specification of the forebrain is regulated by the prechordal plate.

Forebrain development is highly complex and it is clear that a number of genes are implicated in this process. These include *Sonic Hedgehog*, *Fgf8*, *Hesx1*, *HNF3β*, *Nkx2.1*, *Nkx2.2*, *Lim1*, *Otx1*, *Otx2*, *Cerberus*, *goosecoid*, *Pax6*, *Emx1* and *Emx2*. The exact role of many of these genes remains to be defined. Nevertheless, since the forebrain, eyes, optic chiasm, olfactory bulbs, adenohypophysis and hypothalamus originate from the same region in the embryo, defects of some of the genes expressed in the ANR could potentially lead to developmental defects of these structures.

The pituitary gland, which is a midline structure, consists of the anterior, intermediate and posterior lobes, and has a dual embryonic origin. In man, around the fourth week of gestation, a diverticulum called Rathke's pouch projects dorsally from the roof of the stomodeum and grows towards the brain. By the fifth week, the pouch has elongated and become constricted at its attachment to the oral epithelium. By this stage it has come into contact with the infundibulum, a ventral downgrowth from the diencephalon.

Fate map studies in the chick and *Xenopus* have demonstrated that the organ derivatives of the oral roof ectoderm (which include the pituitary and nasal epithelium) are initially derived from the ANR. Upon embryonic turning of the head, the ANR is displaced ventrally and forms the stomodeum, the ectoderm which gives rise to the roof of the mouth and its derived structures [3, 4]. The onset of pituitary organogenesis coincides with a thickening of this initially uniform stomodeal ectoderm on embryonic day (E) 8.5 in mice which then invaginates on E9.0 to form Rathke's pouch. The anterior and intermediate lobes of the pituitary derive from this structure. The posterior lobe is derived as a result of the evagination of the neural ectoderm (infundibulum) at the base of the developing diencephalon in direct contact with Rathke's pouch. Hence, in the mouse, the anterior and intermediate lobes of the pituitary gland form at E8.5 from Rathke's pouch and this makes direct cell-cell contact with the neuroepithelium of the nascent diencephalon at E8.5–9 [5]. The apposition of Rathke's pouch and the diencephalon is maintained throughout the early stages of pituitary organogenesis, and this close relationship together with various animal experimental manipulations has long suggested that inductive tissue interactions are involved in the process. Between E10.5 and E12, the pouch epithelium continues to proliferate as it closes and separates from the underlying oral ectoderm. Subsequent to these initial patterning events, the progenitors of the hormone-secreting cell types proliferate ventrally from the pouch between E12.5 and E15.5 to populate what will form the anterior lobe, which will contain five different cell types [for review, see 6]. The remnants of the dorsal portion of the pouch form the intermediate lobe, containing the melanotrope cell type.

The development of the anterior pituitary gland is dependent on various intrinsic and extrinsic transcription factors and signalling molecules. Extrinsic factors include *Ttf1*, *BMP4* and *Fgf8* [7–9], which are expressed in the diencephalon and not Rathke's pouch, and a role for which has been suggested in normal pituitary development. It has now been established that Rathke's pouch develops in a two-step process which requires at least two sequential inductive signals from the diencephalon. First, the induction and formation of the pouch rudiment is dependent upon BMP4 which is present only in the hypothalamus and not in Rathke's pouch. Secondly, FGF8, which is also present in the hypothalamus and not in Rathke's pouch, activates the key regulatory genes *Lhx3* and *Lhx4* (see below), and these are essential for subsequent development of the pouch rudiment into a definitive pouch [9].

A number of transcription factors are expressed within the pituitary primordium. These include *Six3* [10], *Pax6* [11] and *Hesx1/Rpx* [12, 13], all of which continue to be expressed in Rathke's pouch. Concurrent with organ commitment, two LIM-homeodomain factors, *Lhx3* or *P-Lim* [14, 15] and *Lhx4* [16], are also expressed in Rathke's pouch. Deletion of the *Lhx3/P-Lim* gene in mice results in the failure of the pituitary gland to grow and differentiate, although Rathke's pouch forms [17]. However, a double *Lhx3/Lhx4* mutant results in a failure of formation of Rathke's pouch [16]. The OTX-related factor *Ptx1/P-OTX* is also expressed throughout early pituitary development [18, 19].

The mature anterior pituitary gland is populated by five neuroendocrine cell types, each defined by the hormone produced: corticotropes (ACTH), thyrotropes (TSH), gonadotropes (LH, FSH), somatotropes (GH) and lactotropes (PRL). Each of these cell types differentiates in a temporally and spatially regulated manner [20, 21], and this process is dependent upon a number of transcription factors. A nuclear receptor, SF1, is expressed later in the nascent gonadotrope lineage [22]. A tissue-specific POU domain factor Pit1 is required for terminal differentiation, growth and survival of somatotropes, lactotropes and thyrotropes [23]. Additionally, a paired-like homeodomain factor Prop1 is also required for the determination of the Pit1-dependent lineage [24]. Mutations of both *Pit1* and *Prop1* have been described in man and account for the phenotype of GH, prolactin and TSH deficiency [25, 26] and GH, prolactin, TSH and FSH/LH deficiency [27] respectively. These findings suggest that Prop1 is essential for determination of the gonadotrope lineage as well as the Pit1-dependent lineage in man.

Midline Defects and Pituitary Dysfunction

Congenital midline defects encompass a large array of clinical phenotypes, ranging from those which are incompatible with life to severe palato-facial cleft associated with neuroanatomical defects. The conditions include various forms of holoprosencephaly, septo-optic dysplasia (SOD) and agenesis of the corpus callosum. At the less severe end of the spectrum are isolated cleft lip or palate. The association between hypopituitarism and extensive congenital midline brain defects has long been recognised [28–31]. Rather more controversial is the association between the less severe defects and pituitary dysfunction [32–33]. Of these midline disorders, the commonest is SOD.

Septo-Optic Dysplasia

This is a rare condition which was initially described by Reeves [34] in a 7-month-old baby with absence of the septum pellucidum and optic nerve anomalies, and is often referred to as de Morsier syndrome [35]. It is a highly heterogeneous condition defined loosely by any combination of optic nerve hypoplasia, midline neuroradiological abnormalities such as agenesis of the corpus callosum and absence of the septum pellucidum, and pituitary hypoplasia with consequent panhypopituitarism [34–40]. Other neuroradiological features include cavum septum pellucidum, cerebellar hypoplasia, schizencephaly and aplasia of the fornix [41–43]. The phenotype is highly variable, and each of the three components can occur in isolation or in combination with at least one of the other components. The diagnosis is usually made if two or more of the above features are present. According to Morishima and Aranoff [44], approximately 30% of SOD cases have complete manifestations, 62% of which have the complication of hypopituitarism and 60% have an absent septum pellucidum. De Morsier [35] reported that optic nerve abnormalities were present in only 9 of 36 cases with absent septum pellucidum. Acers [45] studied 45 individuals with optic nerve hypoplasia and found that 12 of these patients had associated midline brain defects, with 6 of these individuals also demonstrating hypopituitarism. Additionally, there appears to be little correlation between the size of the optic nerve and its visual function. The condition is thought to be more frequent in children born to younger mothers (mean maternal age 22 years) [37], although this has been disputed [45]. In some studies [38, 45], there is a preponderance of primigravida mothers.

Optic nerve hypoplasia may be unilateral or bilateral [41, 43, 45], and may be the first presenting feature, with the later onset of endocrine dysfunction [37]. Bilateral optic nerve hypoplasia is commoner (88% as compared with 12% unilateral cases). Neuroradiological abnormalities are present in up to 75–80% of patients with optic nerve hypoplasia [41, 42]. Pituitary hypoplasia may manifest as endocrine deficits varying from isolated growth hormone (GH) deficiency to panhypopituitarism. A decrease in growth rate due to GH deficiency is the commonest feature, with hypoglycaemia and polyuria and polydipsia being less common [37]. Either sexual precocity or failure to develop in puberty may occur [46, 47], and abnormal hypothalamic neuroanatomy or function may be a feature [46, 48–50], as may diabetes insipidus [51]. The endocrinopathy may be evolving with a progressive loss of endocrine function over time [40]. The commonest endocrinopathy is GH deficiency followed by TSH and ACTH deficiency [52, 53]. Gonadotrophin secretion may be retained in the face of other pituitary hormone deficiencies [54]. Commencement of GH treatment in children with GH deficiency may be associated with accelerated pubertal maturation [55]. Neurological deficit is common, and in one study, was documented in 15 of 24 children with a severe degree of optic nerve hypoplasia. The deficit ranged from global retardation to focal deficits such as epilepsy or hemiparesis [56].

Several aetiologies have been postulated to account for the sporadic occurrence of SOD, such as viral infections, environmental teratogens, and vascular or degenerative damage [50, 57]. The syndrome is defined by a developmental malformation of midline forebrain structures, including primarily hypoplasia of the optic nerves and chiasm and agenesis of the septum pellucidum, with other structures such as the corpus callosum, fornix, infundibulum and hypothalamus also being affected. It would appear that the developmental anomaly takes place during a critical period of embryogenesis between 4 and 6 weeks of gestation in man. Two significant developments take place at this stage. First, the telencephalic optic vesicles and retinal ganglion cells differentiate. Secondly, the lamina terminalis thickens, and its subsequent differentiation results in formation of the corpus callosum, anterior commissure and fornix. Any insult which occurs at this stage has the potential to produce failure of ganglion cell formation with subsequent hypoplasia of optic nerves and chiasm in the first instance and lack of commissural or septal formation in the second instance.

Familial cases of SOD are rare, and may be associated with an autosomal recessive inheritance [58–60]. We have recently identified a murine model for SOD [61]. The phenotype of this mouse was the result of targeted disruption of the murine homeobox gene *Hesx1*. Homeobox genes play a crucial role in regional patterning and cell differentiation during the development of multicellular organisms. They were first described in *Drosophila* and are characterized by the presence of a homeobox which encodes a highly conserved 60-amino-acid homeodomain at the protein level. Protein products of these genes can therefore act as specific transcription factors which bind to regulatory elements of target genes by means of the homeodomain. The homeobox gene family can be subdivided into about 30 classes of genes, each characterized by its own specific consensus sequence of the homeodomain [62]. Mutations in several homeobox genes have been identified as the molecular defects in several human anomalies, including *PIT1* and *PROP1* (hypopituitarism and combined hormone deficiencies), *PAX2* (renal-coloboma syndrome), *PAX3* (Waardenberg syndrome type I), *PAX6* (aniridia), *MSX1* (selective tooth agenesis), *MSX2* (craniosynostosis), *HOXD13* (limb deformities), and *SHOX* (short stature) [25–27, 63–71]. The role of homeobox genes in development is elegantly demonstrated by the pituitary gland, and in particular, murine studies have suggested a significant role for *Hesx1* in forebrain and pituitary development.

The Homeobox Gene *Hesx1*

Hesx1 is a member of the paired-like class of homeobox genes [12] and is first expressed during mouse embryogenesis in a small patch of cells in the anterior midline visceral endoderm as gastrulation commences [13, 72, 73]. Expression is then induced in the adjacent ectoderm in an area destined to give rise to the ventral prosencephalon, which will in turn form the forebrain [72]. This ectodermal expression appears to require an inductive signal from the visceral endoderm. Subsequently, *Hesx1* is expressed at the anterior extreme of the rostral neural folds; expression is then restricted to the ventral diencephalon by E9.0. At this stage, *Hesx1* is also expressed in the thickened layer of oral ectoderm which will give rise to Rathke's pouch. *Hesx1* continues to be expressed in the developing anterior pituitary until E12 when its transcripts disappear in a spatiotemporal sequence which corresponds to progressive pituitary cell differentiation. Expression is undetectable by E13.5.

In the Ames dwarf mouse, the phenotype of which arises as a result of a mutation in the *Prop1* gene, *Hesx1* expression persists until E16, suggesting a role for *Prop1* in regulating *Hesx1* expression [27]. Additionally, a role for *Lhx3* in the maintenance of *Hesx1* expression has been suggested by the *Lhx3* null mutants [17]. Since *Hesx1* is a putative repressor gene [61], its role in pituitary development may be to repress downstream target genes. No targets of *Hesx1* have been identified to date.

Targeted disruption of *Hesx1* in the mouse revealed morphological defects in homozygote embryos from E9 and, also, in mutant neonates and adults [61]. These included a reduction in prospective forebrain tissue, absence of developing optic vesicles, markedly decreased head size, craniofacial dysplasia with a short nose and severe microphthalmia. The defects were often asymmetrical, with one side more severely affected than the other. By E10, the severity of the mutant phenotype was variable, and mutants could be divided into either a class I phenotype which was severe or a milder class II phenotype. All homozygotes were abnormal at birth, as were 1% of the heterozygotes. Histological analysis of the mutant mice revealed that class I embryos at E13 had substantially reduced forebrains with no sign of telencephalic vesicle or infundibulum development. Additionally, absence of the optic cups, the olfactory placodes and Rathke's pouch was noted. Class II mutant embryos demonstrated a thin forebrain epithelium with retardation in prosencephalic development. Other abnormalities included reduced telencephalic vesicles with hypothalamic abnormalities and aberrant morphogenesis of Rathke's pouch. The olfactory pits and optic vesicles were also hypoplastic and often asymmetric in appearance.

Further analysis of neonatal and adult mutants revealed hypoplastic nasal cavities, hypoplastic olfactory bulbs, microphthalmia and anophthalmia. Abnormalities of the septum pellucidum and corpus callosum were also noted (fig. 1). The variable forebrain midline defects associated with abnormal morphogenesis of the anterior pituitary gland in the mutant mice were reminiscent of the syndrome of SOD in man [61].

In the light of the phenotype demonstrated in *Hesx1* null mutant mice, we investigated the role of the human homologue of this gene in patients with SOD.

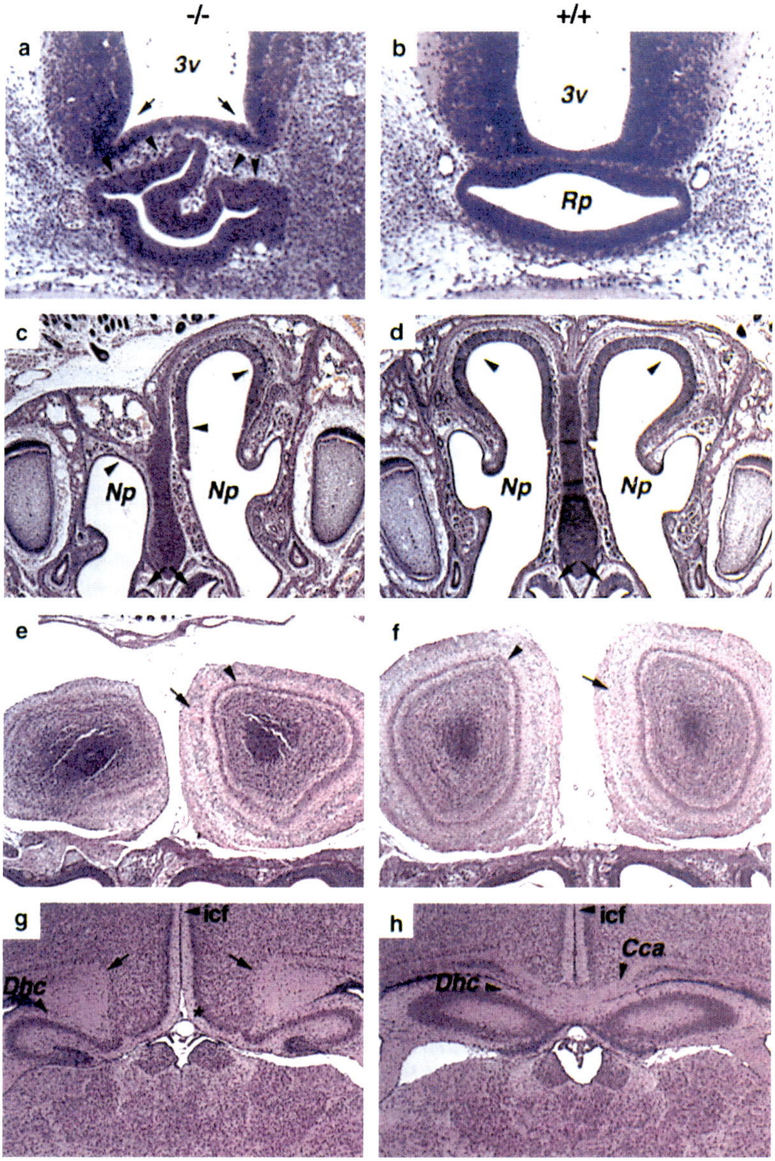

Fig. 1. Histological analysis of the *Hesx1* mutants (from figure 2 in Dattani et al. [61]). ***a***, ***b*** Transverse sections through the hypothalamus and Rathke's pouch of 12.5 dpc (days post-coitum) mutant (***a***) and wild-type (***b***) embryos. Lateral expansion of Rathke's pouch (arrowheads) and bifurcation of the third ventricle (3v; arrows) are evident in the mutant embryo but not in the wild-type littermate. ***c–h*** Transverse sections through the head of mutant (***c***, ***e***, ***g***) and wild-type (***d***, ***f***, ***h***) newborn mice at the levels of the nasal cavities (***c***, ***d***), olfactory bulbs (***e***, ***f***) and anterior hippocampus (***g***, ***h***). ***c***, ***d*** The left nasal passage in the

We isolated the human homologue *HESX1* from a human genomic library. The gene was cloned and sequenced and mapped to chromosome 3p21.1-3p21.2. The *HESX1* locus spans 1.7 kb with a highly conserved genomic organization consisting of four coding exons. The 185-amino-acid open reading frame (ORF) is highly conserved compared with the mouse and *Xenopus* ORFs, particularly across the homeodomains which share 95 and 80% homology respectively. An additional conserved domain is located near the amino-terminus, which is also present in other homeodomain classes including *engrailed* and *goosecoid* and has been postulated to play a role in transcriptional repression [74].

We have now screened 228 patients with a variety of pituitary disorders. Of these, approximately 105 had variations of SOD. Southern blot analysis performed on 38 of these patients revealed no deletions or gross rearrangements associated with the phenotype. Our patient population included a highly consanguineous family in which 2 affected siblings presented with optic nerve hypoplasia, absence of the corpus callosum, and hypoplasia of the pituitary gland with panhypopituitarism [58]. SSCP analysis using primers flanking each of the four *HESX1* coding exons was used to screen these patients for point mutations within the *HESX1* ORF. We found a homozygous missense mutation in the homeobox of *HESX1* in the 2 affected siblings which resulted in the substitution of a highly conserved arginine residue (Arg^{53} in the homeodomain, Arg^{160} in the ORF) by cysteine (R160C). The parents were heterozygous for the mutation. Screening of extended members of the family revealed 9 phenotypically normal heterozygotes within this highly consanguineous pedigree, consistent with an autosomal recessive inheritance. The R160C mutation was not observed in 10 unrelated patients with sporadic SOD or pituitary hypoplasia, nor in 70 normal individuals.

Arg^{160} is one of the most highly conserved residues in homeodomain proteins from a broad range of eukaryotic species. It has been shown to contact the DNA

mutant is atrophic and no nasal neuroepithelium (arrowhead) or vomerulonasal organ (arrow) has developed on this side. The right nasal passage appears normal compared with the wild-type. ***e, f*** The left olfactory bulb is hypoplastic relative to the wild-type and no glomerular (arrows) or mitral (arrowheads) cell layers can be observed on this side. Note the asymmetrical development of both the nasal passages and the olfactory bulbs in the *Hesx1* mutant compared with the wild-type littermate. ***g, h*** Sections through the corpus callosum and dorsal hippocampal commissure of a mutant (***g***) and wild-type brain (***h***). ***g*** 8-day-old mutant neonate showing absence of the corpus callosum, although nerve fibres have formed Probst bundles on either side of the intercerebral fissure (arrows). The dorsal hippocampal commissure also does not cross the midline in the mutant and the intercerebral fissure appears continuous with the third ventricle (*). *Cca* = Corpus callosum; *Dhc* = dorsal hippocampal commissure; *icf* = intercerebral fissure; *Np* = nasal passage; *Rp* = Rathke's pouch; *3v* = third ventricle.

phosphate backbone in the paired- and engrailed-DNA crystal structures [75, 76]. A nonconservative substitution of this amino acid by cysteine would be expected to compromise DNA binding. To test this, recombinant wild-type and mutant HESX1 proteins were purified and their DNA-binding activity assayed in an electromobility shift assay, using the palindromic PIII sequence to which murine Hesx1 has previously been shown to bind [24]. Wild-type HESX1 binds this sequence as a dimer with a high affinity even at concentrations of 1 n*M*, whereas the mutant protein fails to bind even at micromolar concentrations [61]. The severe reduction in in vitro PIII-binding activity of the mutant HESX1 protein compared with the wild-type protein suggests that the R160C substitution may act as a null mutation in vivo.

We have subsequently found five additional heterozygous missense mutations/polymorphisms of *HESX1* in children with various forms of pituitary disease and SOD. Of these, three are nonconservative amino acid substitutions [77]. A nonconservative S170L substitution was identified in 2 siblings with GH deficiency, 1 of whom also demonstrated optic nerve hypoplasia. Haplotype analysis using markers which closely flanked *HESX1* revealed that the 2 siblings had inherited the mutant allele from their unaffected father, whilst both siblings had inherited different maternal haplotypes, thereby excluding compound heterozygosity and suggesting that heterozygous *HESX1* mutations are incompletely penetrant. These data are consistent with our previous finding that *Hesx1*+/– mice display low penetrance of a mild phenotype [61]. Additionally, this mutation was not detected in 100 control chromosomes. The substitution, which was located immediately C-terminal to the homeodomain within an RESQLF motif, which is highly conserved in all HESX1/ANF class proteins, led to a reduction in the DNA-binding affinity of the mutant protein as compared with the wild-type protein. However, this may not be the only alteration in function of the protein. Recent structural studies of the Pbx1 class homeodomains [78, 79] highlight the importance of homeodomain flanking residues/motifs and may provide insight into the functional role of the RESQLF motif. Pbx class proteins, which bind as heterodimers with Hox class proteins, also contain a conserved stretch of residues C-terminal to the homeodomain. In Pbx1 these residues fold into a fourth α-helix which forms an integral part of the homeodomain-binding complex by making specific contacts with homeodomain residues. Hence, the RESQLF motif may have a similar function in that it may mediate interactions with both HESX1 (homodimerization) and other partners (heterodimerization).

We also identified a 541 A-G heterozygous mutation, which results in a nonconservative T181A substitution, in a child with isolated GH deficiency and a hypoplastic anterior pituitary [77]. Again, this mutation was not found in 100 control chromosomes, although the functional significance of this mutation remains unknown. Haplotype analysis again excluded compound heterozygosity since an unaffected heterozygote sibling had inherited the same allele as the index patient from the parent who did not harbour a mutation.

A heterozygous 18 G-C mutation, resulting in a nonconservative Q6H substitution of HESX1, was documented in a child with GH and TSH deficiency [77]. On sequencing exon 1 in 100 control chromosomes, no mutation was found. The functional significance of this mutation remains unknown. In addition to these mutations, we have documented a heterozygous 374 A-G polymorphism in the Afro-Caribbean population, resulting in an N125S substitution in the homeodomain. Finally, we have documented a silent polymorphism in 2 individuals with SOD.

In the heterozygous S170L, T181A and Q6H pedigrees, the inheritance appears to be autosomal dominant with a variable penetrance as was observed in a small proportion of the heterozygous mutant mice. Whether the mechanism by which these mutations lead to a phenotype is due to a dominant negative effect or due to haploinsufficiency remains to be proven.

Holoprosencephaly

Holoprosencephaly is a brain anomaly characterized by a continuum divided into three types: alobar, semilobar and lobar. It is a defect of cleavage, rather than of closure. In the course of normal embryogenesis, the early forebrain (prosencephalon) forms two laterally directed diverticula at 5–6 weeks of gestation. These laterally directed diverticula develop into the paired cerebral hemispheres. Holoprosencephaly may be described as a condition in which the early forebrain (prosencephalon) fails to diverticulate and develops instead into a single unpaired forebrain termed the holoprosencephalon. Because the optic placodes form at the same time and because the olfactory vesicles form about a week later, concurrent nasal and ocular anomalies are common in patients with holoprosencephaly. The alobar type consists of a single ventricle with no separation of the cerebral hemispheres. In the semilobar type, there is fusion of the frontal and parietal lobes and the interhemispheric fissure is present only posteriorly, whilst in the lobar type, the most rostral aspect of the telencephalon, the frontal lobes, are fused ventrally. Absence of the olfactory nerves and bulbs, corpus callosum and the pituitary gland may be associated, as may fusion of the hypothalamus and thalamus [80, 81]. Hence the clinical phenotypes are highly variable. The commonest pituitary abnormality to be associated with holoprosencephaly is diabetes insipidus, but, as with SOD, anterior pituitary hormone deficiencies may also be present.

The incidence is reported as 1 in 16,000 births, but the actual incidence may be higher if all pregnancies are taken into account, since most holoprosencephalies are fatal and would end in spontaneous abortion. The aetiology of holoprosencephaly represents a complex interaction between genetics and the environment. Environmental factors such as in utero exposure to alcohol and retinoic acid have been implicated in the aetiology, as has maternal diabetes mellitus. Several chromosomal disorders such as trisomy 13 are associated with this phenotype. At least 12 chromosomal regions on 11 chromosomes may contain genes involved with holoprosencephaly. The condition is often associated with multiple congenital anomalies, particularly if a cytogenetic abnormality has been identified. Clinically, there is significant variation in phenotypes even among individuals with the identical cytogenetic deletion.

Additionally, autosomal recessive and dominant forms of the disease have also been described. Recently, mutations in four genes have been associated with the phenotype of holoprosencephaly – these are *Sonic Hedgehog (SHH)*, *ZIC2*, *TGIF* and *SIX3*. SHH is a secreted factor expressed early in development in the ventral forebrain and is critical for ventral patterning of the developing neural tube [82, 83]. Homozygous *Shh* null mutant mice displayed abnormalities consistent with holoprosencephaly [84]. Mutations in the coding region of *SHH* have now been documented in a number of sporadic and familial cases of holoprosencephaly [85, 86]. Nevertheless, mutations of *SHH* account for only a minority of familial holoprosencephaly and are a rare cause of sporadic HPE. There is no notable genotype-phenotype correlation between type of mutation or location of the mutation in the SHH gene. In fact, there is a marked intra- and interfamilial variability in the phenotype. A single *SHH* mutation can manifest features from any part of the full spectrum of holoprosencephaly, from alobar to microforms to a clinically normal individual [85, 86].

Brown et al. [87] described mutations in the *ZIC2* (Zinc Finger Protein of Cerebellum, 2) gene in four pedigrees with holoprosencephaly. The heterozygous mutations consisted of insertions and a deletion, resulting in a frameshift in the ORF. The *ZIC2* gene maps to 13q and haploinsufficiency of this gene is likely to account for the brain malformation observed in 13q deletion patients.

Mutations in the homeodomain of *SIX3* have been identified in 4 patients with holoprosencephaly [88]. These mutations were associated with marked phenotypic variability. More recently, missense mutations in the *TGIF* gene, which is expressed during early brain development in the mouse, have also been identified in patients with holoprosencephaly [89]. *TGIF* (TG-interacting factor) is a homeodomain protein which acts as a SMAD2-binding protein and a transcriptional repressor, and competitively inhibits binding of the retinoic acid receptor to a retinoid-responsive promoter. These data suggest that holoprosencephaly is a multigenic disorder, and that the aetiology probably represents a complex interaction between genetics and the environment.

Conclusions

Conditions such as SOD and holoprosencephaly are now known to due to disordered development of the forebrain and related structures. It has previously been speculated that these disorders are due to a complex interaction between various genes and the environment. Mutations in developmental genes have been shown to play a role in the pathogenesis of a number of conditions in man. We have demonstrated a role for the homeobox gene *Hesx1* in a rare familial form of SOD [61]. Our data also suggest a role for this gene in some cases of the more frequently occurring sporadic form of SOD [77]. Nevertheless, the majority of patients screened in our studies show no mutation of *HESX1*. There may be several explanations for this. Alterations in regulatory regions may contribute to the aetiology of this condition. Additionally, like holoprosencephaly, SOD may have a multigenic basis, and mutations in as yet unidentified genes could contribute to some cases. To date, the targets and partners of *Hesx1/HESX1* remain unknown. These may shed further light on the molecular basis of not only SOD, but of various other pituitary disorders. The variability of the phenotypes within a single pedigree in both holoprosencephaly and SOD suggests a complex interaction between genetics and the environment, and at present, the understanding of these conditions is rudimentary. Whatever the mechanism, the striking resemblance between the mouse and human SOD phenotype suggests that *HESX1* may play a similar role in human and mouse embryonic development.

References

1 Beddington RSP, Robertson EJ: Axis development and early asymmetry in mammals. Cell 1999; 96:195–209.

2 Rubenstein JLR, Shimamura K, Martinez S, Puelles L: Regionalization of the prosencephalic neural plate. Annu Rev Neurosci 1998;21:445–477.

3 Couly G, LeDouarin NM: The fate map of the cephalic neural primordium at the presomitic to the 3-somite stage in the avian embryo. Development (Suppl) 1988;103:101–113.

4 Eagleson GW, Harris WA: Mapping of the presumptive brain regions in the neural plate of Xenopus laevis. J Neurobiol 1990,21:427–440.

5 Jacobsen AG, Miyamoto DM, Mai SH: Rathke's pouch morphogenesis in the chick embryo. J Exp Zool 1979;207:351–366.

6 Dasen JS, Rosenfeld MG: Signaling mechanisms in pituitary morphogenesis and cell fate determination. Curr Opin Cell Biol 1999;11:669–677.

7 Kimura S, Hara Y, Pineau T, Fernandez-Salguero P, Fox CH, Ward JM, Gonzalez FJ: The T/ebp null mouse: Thyroid-specific enhancer-binding protein is essential for the organogenesis of the thyroid, lung, ventral forebrain, and pituitary. Genes Dev 1996;10:60–69.

8 Treier M, Gleiberman AS, O'Connell SM, Szeto DP, McMahon JA, McMahon AP, Rosenfeld MG: Multistep signaling requirements for pituitary organogenesis *in vivo*. Genes Dev 1998;12:1691–1704.

9 Takuma N, Sheng HZ, Furuta Y, Ward JM, Sharma K, Hogan BLM, Pfaff SL, Westphal H, Kimura S, Mahon KA: Formation of Rathke's pouch requires dual induction from the diencephalon. Development 1998;125:4835–4840.

10 Oliver G, Wehr R, Jenkins NA, Copeland NG, Cheyette BN, Hartenstein V, Zipursky SL, Gruss P: Homeobox genes and connective tissue patterning. Development 1995;12:693–705.

11 Walther C, Gruss P: *Pax6*, a murine paired box gene, is expressed in the developing CNS. Development 1991;113:1435–1449.

12 Thomas PQ, Johnson BV, Rathjen J, Rathjen PD: Sequence, genomic organization, and expression of the novel homeobox gene *Hesx1*. J Biol Chem 1995;270:3869–3875.

13 Hermesz E, Mackem S, Mahon KA: *Rpx*: A novel anterior-restricted homeobox gene progressively activated in the prechordal plate, anterior neural plate and Rathke's pouch of the mouse embryo. Development 1996;122:41–52.

14 Zhadanov AB, Bertuzzi S, Taira M, Dawid IB, Westphal H: Expression pattern of the murine LIM class homeobox gene Lhx3 in subsets of neural and neuroendocrine tissues. Dev Dyn 1995;202: 354–364.

15 Bach I, Rhodes SJ, Pearse RV, Heinzel T, Gloss B, Scully KM, Sawchenko PE, Rosenfeld MG: *P-Lim*, a LIM homeodomain factor, is expressed during pituitary organ and cell commitment and synergizes with *Pit1*. Proc Natl Acad Sci USA 1995;92:2720–2724.

16 Sheng HZ, Moriyama K, Yamashita T, Li H, Potter SS, Mahon KA, Westphal H: Multistep control of pituitary organogenesis. Science 1997;278:1809–1812.

17 Sheng HZ, Zhadanov AB, Mosinger B Jr, Fujii T, Bertuzzi S, Grinberg A, Lee EJ, Huang SP, Mahon KA, Westphal H: Specification of pituitary cell lineages by the LIM homeobox gene *Lhx3*. Science 1996;272:1004–1007.

18 Lamonerie T, Tremblay JJ, Lanctot C, Therrien Y, Gauthier Y, Drouin J: Ptx1, a *bicoid*-related homeobox transcription factor involved in transcription of the proopiomelanocortin gene. Genes Dev 1996;10:1284–1295.

19 Szeto DP, Ryan AK, O'Connell SM, Rosenfeld MG: P-OTX: A Pit1 interacting homeodomain factor expressed during anterior pituitary gland development. Proc Natl Acad Sci USA 1996;93: 7706–7710.

20 Simmons DM, Voss JW, Ingraham HA, Holloway JM, Broide RS, Rosenfeld MG, Swanson LW: Pituitary cell phenotypes involve cell-specific Pit-1 mRNA translation and synergistic interactions with other classes of transcription factors. Genes Dev 1990;4:695–711.

21 Japon MA, Rubinstein M, Low MJ: In situ hybridization analysis of anterior pituitary hormone gene expression during fetal mouse development. J Histochem Cytochem 1994;42:1117–1125.

22 Ingraham HA, Lala DS, Ikeda Y, Luo X, Shen WH, Nachtigal MW, et al: The nuclear receptor steroidogenic factor-1 acts at multiple levels of the reproductive axis. Genes Dev 1994;8:2302–2312.

23 Li S, Crenshaw EBI, Rawson EJ, Simmons DM, Swanson LW, Rosenfeld MG: Dwarf locus mutants which lack three pituitary cell types result from mutations in the POU domain gene Pit1. Nature 1990;347:528–533.

24 Sornson MW, Wu W, Dasen JS, Flynn SE, Norman DJ, O'Connell SM, Gukovsky I, Carriere C, Ryan AK, Miller AP, Zuo L, Gleiberman AS, Andersen B, Beamer WG, Rosenfeld MG: Pituitary lineage determination by the Prophet of Pit-1 homeodomain factor defective in Ames dwarfism. Nature 1996;384:327–333.

25 Pfäffle RW, DiMattia GE, Parks JS, Brown MR, Wit JM, Jansen M, Van der Nat H, Van den Brande JL, Rosenfeld MG, Ingraham HA: Mutation of the POU-specific domain of Pit-1 and hypopituitarism without pituitary hypoplasia. Science 1992;257:1118–1121.

26 Radovick S, Nations M, Du Y, Berg LA, Weintraub BD, Wondisford FE: A mutation in the POU-homeodomain of Pit-1 responsible for combined pituitary hormone deficiency. Science 1992;257: 1115–1117.

27 Wu W, Cogan JD, Pfäffle RW, Dasen JS, Frisch H, O'Connell MG, Flynn SE, Brown MR, Mullis PE, Parks JS, Phillips JA III, Rosenfeld MG: Mutations in *PROP1* cause familial combined pituitary hormone deficiency. Nat Genet 1998;18:147–149.

28 Artman HG, Boyden E: Microphthalmia with a single central incisor and hypopituitarism. J Med Genet 1990;27:192–193.

29 Brook CGD, Sanders MD, Hoare RD: Septo-optic dysplasia. Br Med J 1972;iii:811–813.

30 Coulter CL, Leech RW, Schaefer GB, Scheithauer BW, Brumback RA: Midline cerebral dysgenesis, dysfunction of the hypothalamo-pituitary axis and fetal alcohol effects. Arch Neurol 1993;50:771–775.
31 Fitz CR: Holoprosencephaly and septo-optic dysplasia. Neuroimaging Clin N Am 1994;4:263–281.
32 Lee J, Nunn J, Wright C: Height and weight achievement in cleft lip and palate. Arch Dis Child 1997;76:70–72.
33 Rudman D, Davis GT, Priest JH, Patterson JH, Kutner MH, Heymsfield SB, Bethel RA: Prevalence of growth hormone deficiency in children with cleft lip or palate. J Pediatr 1978;93:378–382.
34 Reeves DL: Congenital absence of septum pellucidum. Bull Johns Hopkins Hosp 1941;69:61–71.
35 De Morsier G: Etudes sur les dysraphies cranio-encéphaliques. III. Agénésie du septum lucidum avec malformation du tractus optique: La dysplasie septo-optique. Schweiz Arch Neurol Psychiatr 1956;77:267–292.
36 Hoyt WF, Kaplan SL, Grumbach MM, Glaser JS: Septo-optic dysplasia and pituitary dwarfism. Lancet 1970;i:893–894.
37 Arslanian SA: Hormonal, metabolic and neuroradiologic abnormalities associated with septo-optic dysplasia. Acta Endocrinol 1984;107:282–288.
38 Izenberg N, Rosenblum M, Parks JS: The endocrine spectrum of septo-optic dysplasia. Clin Pediatr 1984;23:632–636.
39 Roessman U: Septo-optic dysplasia or de Morsier syndrome. J Clin Neuroophthalmol 1989;9:156–159.
40 Stanhope R, Preece MA, Brook CGD: Hypoplastic optic nerves and pituitary dysfunction. A spectrum of anatomical and endocrine abnormalities. Arch Dis Child 1984;59:111–114.
41 Brodsky MC, Glasier CM: Optic nerve hypoplasia: Clinical significance of associated central nervous system abnormalities on magnetic resonance imaging. Arch Ophthalmol 1993;111:66–73.
42 Zeki SM, Hollman AS, Dutton GN: Neuroradiological features of patients with optic nerve hypoplasia. J Pediatr Ophthalmol Strabismus 1992;29:107–111.
43 Willnow S, Kiess W, Butenandt O, Dorr HG, Enders A, Strasser-Vogel B, Egger J, Schwarz HP: Endocrine disorders in septo-optic dysplasia (de Morsier syndrome) – Evaluation and follow-up of 18 patients. Eur J Pediatr 1996;155:179–184.
44 Morishima A, Aranoff GS: Syndrome of septo-optic pituitary dysplasia: The clinical spectrum. Brain Dev 1986;8:233–239.
45 Acers TE: Optic nerve hypoplasia: Septo-optic-pituitary dysplasia syndrome. Trans Am Ophthalmol Soc 1981;79:425–427.
46 Huseman CA, Kelch RP, Hopwood NJ, Zipf WB: Sexual precocity in association with septo-optic dysplasia and hypothalamic hypopituitarism. J Pediatr 1978;92:748–753.
47 Hanna CE, Mandel SH, LaFranchi SH: Puberty in the syndrome of septo-optic dysplasia. Am J Dis Child 1989;43:186–189.
48 Lam KSL, Wang C, Ma JT, Leung SP, Yeung RT: Hypothalamic defects in two adult patients with septo-optic dysplasia. Acta Endocrinol 1986;112:305–309.
49 Yukizane S, Kimura Y, Yamashita Y, Matsuishi T, Horikawa H, Ando H, Yamashita F: Growth hormone deficiency of hypothalamic origin in septo-optic dysplasia. Eur J Paediatr 1990;150:30–33.
50 Roesmann U, Velasco ME, Small EJ, Hori A: Neuropathology of 'septo-optic dysplasia' (de Morsier syndrome) with immunohistochemical studies of the hypothalamus and pituitary gland. J Neuropathol Exp Neurol 1987;64:597–608.
51 Masera N, Grant DB, Stanhope RG, Preece MA: Diabetes insipidus with impaired osmotic regulation in septo-optic dysplasia and agenesis of the corpus callosum. Arch Dis Child 1994;70:51–53.
52 Cameron FJ, Khadilkar V, Stanhope R: Pituitary dysfunction, morbidity and mortality with congenital midline malformation of the cerebrum. Eur J Pediatr 1999;158:97–102.
53 Costin G, Murphree AL: Hypothalamic-pituitary function in children with optic nerve hypoplasia. Am J Dis Child 1985;139:249–254.
54 Nanduri VR, Stanhope R: Why is the retention of gonadotrophin secretion common in children with panhypopituitarism due to septo-optic dysplasia? Eur J Endocrinol 1999;140:48–50.

55 Freude S, Frisch H, Wimberger D, Schober E, Hausler E, Waldhauser F, Aichner F: Septo-optic dysplasia and growth hormone deficiency: Accelerated pubertal maturation during GH therapy. Acta Paediatr 1992;81:641–645.

56 Roberts-Harry J, Green SH, Willshaw HE: Optic nerve hypoplasia: Associations and management. Arch Dis Child 1990;65:103–106.

57 Zaias B, Becker D: Septo-optic dysplasia: Developmental or acquired abnormality? A case report. Trans Am Neurol Assoc 1978;103:273–277.

58 Wales JKH, Quarrell OWJ: Evidence for possible Mendelian inheritance of septo-optic dysplasia. Acta Pediatr 1996;85:391–392.

59 Blethen SL, Weldon VV: Hypopituitarism and septo-optic dysplasia in first cousins. Am J Med Genet 1985;21:123–129.

60 Benner JD, Preslan MW, Gratz E, Joslyn J, Schwartz M, Kelman S: Septo-optic dysplasia in two siblings. Am J Ophthalmol 1990;109:632–637.

61 Dattani MT, Martinez-Barbera JP, Thomas PQ, Brickman JM, Gupta R, Martensson I, Toresson H, Fox M, Wales JKH, Hindmarsh PC, Krauss S, Beddington RSP, Robinson ICAF: Mutations in the homeobox gene *HESX1/Hesx1* associated with septo-optic dysplasia in human and mouse. Nat Genet 1998;19:125–133.

62 Kappen C, Schughart K, Ruddle FH: Early evolutionary origin of major homeodomain sequence classes. Genomics 1993;18:54–70.

63 Sanyanusin P, Schimmenti LA, McNoe LA, Ward TA, Pierpont MEM, Sullivan MJ, Dobyns WB, Eccles MR: Mutation of the *PAX2* gene in a family with optic nerve colobomas, renal anomalies and vesicoureteral reflux. Nat Genet 1995;9:358–363.

64 Baldwin CT, Hoth CF, Amos JA, da Silva EO, Milunsky A: An exonic mutation in the *HuP2* paired domain gene causes Waardenburg's syndrome. Nature 1992;355:637–638.

65 Morell R, Friedman TB, Asher JH Jr: A plus-one frameshift mutation in *PAX3* alters the entire deduced amino-acid sequence of the paired box in a Waardenburg syndrome type 1 (WS1) family. Hum Mol Genet 1993;2:1487–1488.

66 Tassabehji M, Read AP, Newton VE, Harris R, Balling R, Gruss P, Strachan T: Waardenburg's syndrome patients have mutations in the human homologue of the *Pax-3* paired box gene. Nature 1992;355:635–636.

67 Glaser T, Walton DS, Maas RL: Genomic structure, evolutionary conservation and aniridia mutations in the human *PAX6* gene. Nat Genet 1992;2:232–238.

68 Vastardis H, Karimbux N, Guthua SW, Seidman JG, Seidman CE: A human *MSX1* homeodomain missense mutation causes selective tooth agenesis. Nat Genet 1996;13:417–421.

69 Jabs EW, Muller U, Li X, Ma L, Luo W, Haworth IS, Klisak I, Sparkes R, Warman ML, Mulliken JB, Snead ML, Maxson R: A mutation in the homeodomain of the human *MSX2* gene in a family affected with autosomal dominant craniosynostosis. Cell 1993;75:443–450.

70 Muragaki Y, Mundlos S, Upton J, Olsen BR: Altered growth and branching patterns in synpolydactyly caused by mutations in *HOXD13*. Science 1996;272:548–551.

71 Rao E, Weiss B, Fukami M, Rump A, Niesler B, Mertz A, Muroya K, Binder G, Kirsch S, Winkelmann M, Nordsiek G, Heinrich U, Breuning MH, Ranke MB, Rosenthal A, Ogata T, Rappold GA: Pseudoautosomal deletions encompassing a novel homeobox gene cause growth failure in idiopathic short stature and Turner syndrome. Nat Genet 1997;16:147–149.

72 Thomas P, Beddington RSP: Anterior primitive endoderm may be responsible for patterning the anterior neural plate in the mouse embryo. Curr Biol 1996;6:1487–1496.

73 Thomas P, Brickman J, Popperl H, Krumlauf R, Beddington RSP: Axis duplication and anterior identity in the mouse embryo; in Symposia on Quantitative Biology. Cold Spring Harbor, Cold Spring Harbor Laboratory Press, 1997, vol 62, pp 115–125.

74 Smith ST, Jaynes JB: A conserved region of engrailed, shared among all en-, gsc-, Nk1-, Nk2- and msh-class homeoproteins, mediates active transcriptional repression in vivo. Development 1996; 122:3141–3150.

75 Wilson DS, Guenther B, Desplan C, Kuriyan J: High-resolution crystal structure of a paired (Pax) class cooperative homeodomain dimer on DNA. Cell 1995;82:709–719.

76 Kissinger CR, Lui B, Martin-Blanco E, Kornberg TB, Pabo CO: Crystal structure of an engrailed homeodomain-DNA complex at 2.8 Å resolution: A framework for understanding homeodomain-DNA interactions. Cell 1990;63:579–590.
77 Thomas PQ, Dattani MT, Brickman JM, McNay D, Warne G, Zacharin M, Cameron F, Hurst J, Woods K, Dunger D, Stanhope R, Forrest S, Robinson ICAF, Beddington RSP: Heterozygous *HESX1* mutations associated with isolated congenital pituitary hypoplasia and septo-optic dysplasia. Hum Mol Genet 2001;10:39–45.
78 Piper DE, Batchelor AH, Chang CP, Cleary ML, Wolberger C: Structure of a HoxB1-Pbx1 heterodimer bound to DNA: Role of the hexapeptide and a fourth homeodomain helix in complex formation. Cell 1999;96:587–597.
79 Lu Q, Kamps MP: Structural determinants within Pbx1 that mediate cooperative DNA binding with pentapeptide-containing Hox proteins: Proposal for a model of a Pbx1-Hox-DNA complex. Mol Cell Biol 1996;16:1632–1640.
80 Fitz CR: Holoprosencephaly and related entities. Neuroradiology 1983;25:225–238.
81 Fitz CR: Holoprosencephaly and septo-optic dysplasia. Pediatr Neuroradiol 1994;4:263–281.
82 Echelard Y, Epstein DJ, St Jacques B, Shen L, Mohler J, McMahon JA, McMahon AP: Sonic Hedgehog, a member of a family of putative signalling molecules, is implicated in the regulation of CNS polarity. Cell 1993;75:1417–1430.
83 Roelink H, Augsburger A, Heemskerk J, Korzh V, Norlin S, Ruiz I Altaba A, Tanabe Y, Placzek M, Edlund T, Jessell TM, Dodd J: Floor plate and motor neuron induction by vhh-1, a vertebrate homolog of hedgehog expressed by the notochord. Cell 1994;76:761–775.
84 Chiang C, Litingtung Y, Lee E, Young KE, Corden JL, Westphal H, Beachy PA: Cyclopia and defective axial patterning in mice lacking Sonic Hedgehog gene function. Nature 1996;383:407–413.
85 Roessler E, Belloni E, Gaudenz K, Jay P, Berta P, Scherer SW, Tsui LC, Muenke M: Mutations in the human Sonic Hedgehog gene cause holoprosencephaly. Nat Genet 1996;14:357–360.
86 Roessler E, Belloni E, Gaudenz K, Vargas F, Scherer SW, Tsui LC, Muenke M: Mutations in the carboxy-terminal domain of Sonic Hedgehog gene cause holoprosencephaly. Hum Mol Genet 1997; 6:1847–1853.
87 Brown SA, Warburton D, Brown LY, Yu C, Roeder ER, Stengel-Rutkowski S, Hennekam RCM, Muenke M: Holoprosencephaly due to mutations in *ZIC2*, a homologue of *Drosophila* odd-paired. Nat Genet 1998;20:180–183.
88 Wallis DE, Roessler E, Hehr U, Nanni L, Wiltshire T, Richieri-Costa A, Gillessen-Kaesbach G, Zackai EH, Rommens J, Muenke M: Mutations in the homeodomain of the human *SIX3* gene cause holoprosencephaly. Nat Genet 1999;22:196–198.
89 Gripp KW, Wotton D, Edwards MC, Roessler E, Ades L, Meinecke P, Richieri-Costa A, Zackai EH, Massague J, Muenke M, Elledge SJ: Mutations in TGIF cause holoprosencephaly and link NODAL signalling to human neural axis determination. Nat Genet 2000;25:205–208.

M.T. Dattani, MBBS, FRCPCH, FRCP, MD, London Centre for Paediatric Endocrinology
Institute of Child Health, 30 Guilford Street, London WC1N 1EH (UK)
Tel. +44 171 242 9789, Fax +44 171 404 6191

Rappaport R, Amselem S (eds): Hypothalamic-Pituitary Development. Genetic and Clinical Aspects.
Endocr Dev. Basel, Karger, 2001, vol 4, pp 94–108

Magnetic Resonance Imaging of the Hypothalamic-Pituitary Region in Nontumoral Hypopituitarism

Irène Netchine[a], *Juliane Léger*[b], *Raphaël Rappaport*[c]

[a] Laboratoire d'Explorations Endocriennes, Hôpital Troussseau, Paris et Institut National de la Santé et de la Recherche Médicale (Unité 468), Hôpital Henri-Mondor, Créteil;
[b] Service d'Endocrinologie et Diabetologie, Hôpital Robert-Debré, Paris et
[c] Fédération de Pédiatrie, Hôpital Necker-Enfant Malades, Paris, France

Introduction

Growth hormone deficiency (GHD) may be either isolated (isolated growth hormone deficiency, IGHD) or associated with a deficit of at least one other anterior pituitary hormone (combined pituitary hormone deficiency, CPHD). Furthermore, IGHD may evolve towards CPHD. Such forms of nontumoral hypopituitarism may be either isolated or associated with various congenital midline defects. Within this rare condition, a small number of cases are familial, with either an autosomal dominant, recessive or X-linked mode of inheritance, suggesting the implication of various genetic abnormalities in nontumoral hypopituitarism [1]. This condition is therefore highly heterogeneous in terms of clinical presentation, natural history of the disease phenotype, hormonal phenotype, family history and may be associated with extrapituitary congenital developmental defects [2]. When GHD is isolated, except for complete deficiency, it is often difficult to establish a diagnosis on the basis of laboratory work-up alone (i.e. GH peak response to pharmacological stimulation tests, IGF1, IGF1-BP3) [3]. The use of magnetic resonance imaging (MRI) in the work-up of suspected hypothalamic-pituitary disorders has proven to be a useful tool for establishing the diagnosis of hypopituitarism [4]. The various developmental defects of the hypothalamic-pituitary region recognized by MRI are also key elements in guiding the choice of the candidate gene to be investigated in hypopituitarism [5]. We will review the main findings from MRI studies of the hypothalamic-pituitary region performed in patients with hypopituitarism and discuss their consequences in terms of diagnosis and etiopathogenic implications.

Hypothalamic-Pituitary Region in Normal Subjects

Normal Development and Structure of the Hypothalamic-Pituitary Region

The pituitary gland is linked to the median eminence by the pituitary stalk and consists of the anterior, intermediate and posterior lobes. Its development occurs early in all species and takes place between the fourth and ninth weeks of gestation in humans [6]. The anterior lobe originates from a midline region located immediately anterior to the neural plate. The hypothalamus and the posterior pituitary stem from the neural plate itself. The posterior lobe, or neurohypophysis, originates as a downward extension, called infundibulum, from the hypothalamus. The primordium of the anterior pituitary, called Rathke's pouch, projects dorsally from the roof of the stomodeum and grows towards the infundibulum. In the same period, the distal portion of the infundibulum differentiates to form the posterior pituitary. The co-development of these two structures is the result of a temporospatial cascade of molecular events with activation and repression of numerous transcription factors [7]. The pituitary stalk consists primarily of the neural connection between the median eminence and the posterior pituitary lobe. This neural connection consists of axonal structures in which vasopressin and oxytocin travel to the posterior lobe, from which they are released by exocytosis. The pituitary stalk also contains a delicate layer of tissue that is permeated by numerous capillary loops of the hypophyseal-portal blood system. This vascular structure also provides the principal blood supply to the anterior pituitary as there is little direct arterial supply to this organ. In contrast, the posterior pituitary has a direct vascular supply.

MRI of the Hypothalamic-Pituitary Region in Normal Subjects

MRI provides a precise anatomical evaluation of the hypothalamic-pituitary region and its surrounding structures. Figure 1 depicts a normal hypothalamic-pituitary region on sagittal (fig. 1A) and coronal (fig. 1B) views. A MRI work-up of the hypothalamic-pituitary region should include sagittal and coronal views of the pituitary area using a T1-weighted sequence before and after contrast enhancement. Sagittal slices allow to evaluate the integrity of the corpus callosum. A dynamic series of coronal slices acquired at the early phase of contrast enhancement may reveal heterogeneities within the gland which may not be detected otherwise. A coronal T2-weighted sequence may be useful in detecting cystic com-

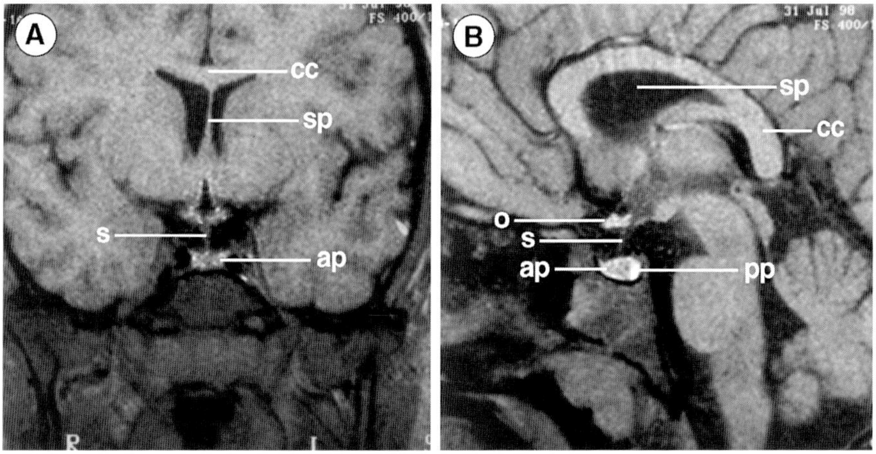

Fig. 1. Normal hypothalamic-pituitary region (*A* sagittal view, *B* coronal view). ap = Anterior pituitary, pp = posterior pituitary, s = pituitary stalk, o = optic chiasm, cc = corpus callosum, sp = septum pellucidum.

ponents of the gland or in revealing other heterogeneities. These slices should include the optic pathways. An axial T2-weighted sequence covering the whole brain provides a view of the septum lucidum and the cerebral cortex. Those slices, which focus on the hypothalamic-pituitary region, should be no thicker than 3 mm to avoid partial volume effects which can lead to errors in evaluating small structures. The maximum height of the anterior pituitary gland is measured in a plane perpendicular to the floor of the sella turcica on sagittal T1-weighted image [29]. The mean anterior pituitary height varies according to age. In neonates the mean height is 4.5 mm, as the pituitary is typically convex upward. After age of 2 months, the superior aspect flattens and the height decreases until puberty, at which time there is a dramatic increase in pituitary height. Although very large series of published normative data of anterior pituitary height in children are not available, during childhood, a mean anterior pituitary height of 2 mm is considered to be below –2 SD [9, 10]. The volume of the anterior pituitary may also be measured, but the procedure requires time-consuming post-processing of the data and is not available in all institutions [4, 11]. The posterior pituitary lobe is normally a round or oval structure of high signal intensity in the posterior part of the sella on T1-weighted images. Although it is still controversial, it is believed that the phospholipid component of the neurohypophysis accounts for the spontaneous T1-weighted brightness of the posterior pituitary. This brightness is related to the functional integrity of the posterior pituitary. However, an absence of high

Table 1. Reports from the literature of MRI findings in patients (n > 35) with hypopituitarism

Reference	Year	Patients	MRI image			Presenting pituitary function	
			EPP	HA	NA	IGHD	CPHD
Bressami et al. [16]	1990	57	34 (59%)	23 (40%)	0	36 (EPP = 37%)	21 (EPP = 95%)
Maghnie et al. [20]	1991	45	20 (44%)	13 (29%)	12 (26%)	33 (EPP = 24%)	8 (EPP = 100%)
Agyropoulou et al. [23]	1992	46	29 (63%)	10 (21%)	7 (15%)	26 (EPP = 38%)	20 (EPP = 95%)
Triulzi et al. [4]	1994	101	59 (48%)	UD	UD	67 (EPP = 44%)	34 (EPP = 85%)
Nagel et al. [25]	1997	56	24 (42%)	11 (19%)	17 (30%)	43 (EPP = 27%)	13 (EPP = 92%)
Kornreich et al. [10]	1997	51	41 (80%)	0	10	23 (EPP = 56%)	28 (EPP = 100%)
Kornreich et al. [27]	1998	44	41 (93%)	UD	UD	21 (EPP = 81%)	23 (EPP = 91%)
Hamilton et al. [29]	1998	35	25 (71%)	3 (<1%)	2 (<1%)	20 (EPP = 70%)	15 (EPP = 73%)
Arifa et al. [30]	1999	100	20 (20%)	28 (28%)	41 (41%)	79 (EPP = UD)	19 (EPP = UD)

EPP = Ectopic posterior pituitary; HA = hypoplastic anterior pituitary with normal located posterior pituitary; NA = normal anatomy of the hypothalamic-pituitary region with normal height of the anterior pituitary; IGHD = isolated growth hormone deficiency; CPHD = combined pituitary hormone deficiency; UD = undetermined.

intensity signal may rarely be seen in subjects with a normal posterior pituitary function in older adult patients [12]. The posterior pituitary is easily distinguished from the bright signal of fat within the marrow of the dorsum and clivus and should be distinguished from a stalk-located lipoma [13]. The pituitary stalk is normally visualized without contrast enhancement. When this is not the case, contrast enhancement may show the vascular components of the pituitary stalk and the use of dynamic MRI sequences may provide additional information concerning the partial preservation of the hypothalamic-hypophyseal portal system [14].

MRI Findings in Patients with GHD

MRI performed in patients with nontumoral hypopituitarism has revealed a variety of anatomical abnormalities of the hypothalamic-pituitary region [4, 10, 15–31]. A characteristic finding, known as 'pituitary stalk interruption syndrome' (PSIS), combines ectopic posterior pituitary, interrupted pituitary stalk and anterior pituitary hypoplasia. Another large group of patients show only isolated anterior pituitary hypoplasia with a posterior pituitary located in its normal anatomical position. The respective frequency of these MRI findings in patients with hypopituitarism and the distribution between IGHD and CPHD has varied among published reports (table 1). Strikingly, some cases of enlarged sella turcica with hyperplasia or hypoplasia of the anterior pituitary have also been described

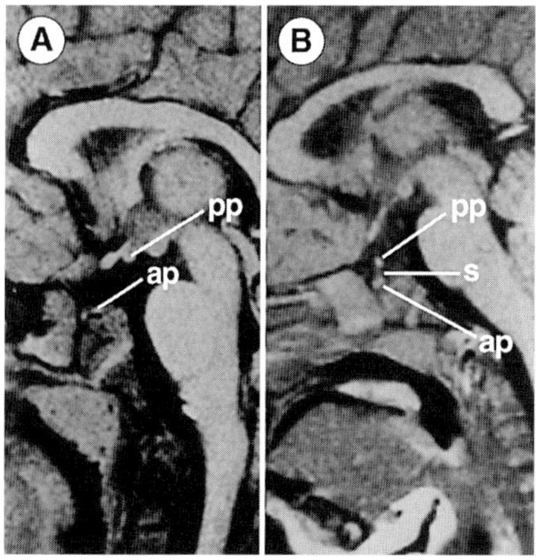

Fig. 2. Variable position of the ectopic posterior pituitary (***A*** in the median eminence of the hypothalamus, ***B*** in the pituitary stalk). ap = Anterior pituitary, pp = posterior pituitary, s = pituitary stalk.

in patients with congenital hypopituitarism [5, 32, 33]. Finally, some patients evaluated because of IGHD show no structural abnormalities of the hypothalamic-pituitary region.

Ectopic Posterior Pituitary, Abnormal Pituitary Stalk and Anterior Pituitary Hypoplasia Triad

Ectopic posterior pituitary (EPP) is defined by a characteristic MRI signal intensity of the posterior lobe of the pituitary in the region of the median eminence of the hypothalamus and absence of normal signal in the usual position of the posterior pituitary lobe (fig. 2A, 3). It was initially described as part of a typical triad also including lack of visibility of the pituitary stalk (hence considered as interrupted) and anterior pituitary hypoplasia [34, 35]. Although the ectopic bright signal remains the most precisely identifiable part of this syndrome, it may have variable positions: it has recently been shown that the EPP bright signal could be located at different levels (proximal, middle or distal) along a visible pituitary stalk (fig. 2A, B) [27, 30].

On the other hand, careful imaging of the pituitary stalk has demonstrated that it may be either absent, thin, normal or even enlarged [36]. It has been shown that the sensitivity of MRI for visualizing the pituitary stalk is increased by contrast enhancement which can show the presence of an extremely thin pituitary stalk [14, 31, 37]. Therefore, it seems that a residual hypothalamo-pituitary connection may be preserved. A sensitive test based on the dopaminergic control of prolactin secretion has been used in patients with congenital GHD: domperidone, a receptor-directed antidopaminergic drug inducing a prolactin increase in normal subjects, elicited a reduced but persistent response in patients with EPP [38]. It was concluded that a residual hypothalamo-hypophyseal connection was preserved even if the pituitary stalk could not be identified by MRI. In a recent study, the cases combining EPP with a visible pituitary stalk represented half of the patients with EPP. Among them, 50% of these latter cases showed a normal anterior lobe. In contrast, the anterior pituitary was hypoplastic in the vast majority of patients showing no visibility of the pituitary stalk [31].

CPHD has been shown to be associated with EPP on MRI at a high rate ranging from 73 to 100% of the cases in the meta-analysis presented in table 1 [4, 10, 16, 20, 23, 25, 27, 29, 30]. However, IGHD could also be the presenting symptom for EPP. In this subgroup of patients a visible pituitary stalk was more frequently found. Indeed this combination may indicate a better functional prognosis of the anterior pituitary function with a lower risk of developing other pituitary deficiencies at a later age [14, 31]. A careful follow-up of these patients is required with reassessment of the anterior pituitary functions. Presently, the natural history of the disease is not well established. In our experience the mean growth retardation at diagnosis was identical whether patients had IGHD or CPHD [31]. In most cases, growth failure occurs in early childhood. Alternatively, it may be delayed and height may remain at –2 SD until the age of puberty. Diagnosis of defective hypothalamic-pituitary connection with EPP has rarely been made in other conditions such as neonatal secondary hypothyroidism with low TSH or delayed puberty with gonadotropin deficiency. Importantly, with few exceptions, there is no diabetes insipidus.

A high prevalence of associated malformations has been reported in patients with PSIS. Abnormalities of the skull and brain are the most frequent; Arnold Chiari malformation (both grade I and II), agenesis of corpus callosum, persistence of the craniopharyngeal canal, aplasia of sella turcica or septo-optic dysplasia (fig. 3) were present in 75% of these cases [31]. Other noncerebral associated malformations, mostly classified as midline abnormalities, have also been described. They include palatine cleft, single median incisor, nasal pyriform aperture stenosis, coloboma, anus anteposition, hand atrophy, interatrial septal defect, persistent common atriventricular ostrium, single ventricle and situs inversus [2, 4, 30, 31, 39–41].

The pathogenesis of PSIS is unknown. Perinatal trauma was reported in some patients but did not appear to be consistently associated with PSIS. Alternatively, a developmental defect occurring prenatally is suggested by many lines of evidence: (1) the frequent association with other congenital abnormalities; (2) the lack of significant frequency of birth trauma; (3) the occurrence of an increased risk of breach delivery rather interpreted as a consequence of prenatal

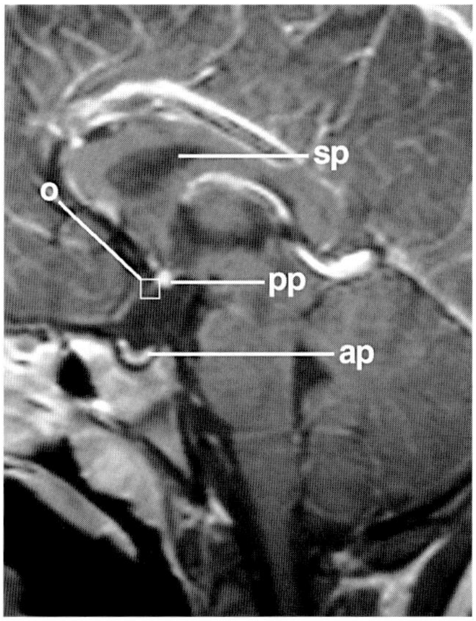

Fig. 3. Septo-optic dysplasia associated with EPP, hypoplastic anterior pituitary (ap) and absence of pituitary stalk. Hypoplasia of the optic chiasm (o) with a normal septum pellucidum (sp).

hypopituitarism; (4) the report of hypopituitary twins with a paracentric inversion of the short arm of chromosome 1: both patients died before the age of 2 months (postmortem examination revealed that the anterior pituitary was small and located within the sella while the posterior pituitary was most probably located at the base of the median eminence; there was no remnant of a pituitary stalk, no portal blood vessels, no hemorrhage or fibrosis and no evidence of any other reaction to traumatic interruption of the stalk [42]); (5) the existence of familial cases of congenital GHD with EPP [31, 40]. In addition, a genetic defect may have occurred in the very few reported familial cases as well as those (unpublished) shown in figure 4, suggesting different transmission patterns [8, 23, 31, 40, 42]. Furthermore within the same family, different phenotypes may be present as reported in 2 sibs (fig. 2) – 1 showing an EPP hyperintense signal located in the median eminence and absence of pituitary stalk visibility with CPHD (patient 1, fig. 2A), the other having an EPP located in the middle of a visible pituitary stalk with IGHD (patient 2, fig. 2B) [30]. Hence, distinct clinical presentations may likely result from a common early defect in the Rathke's pouch primordium of the anterior lobe and the distal portion of the infundibulum forming the posterior pituitary. Possibly a disruptive vascular event may have compromised the normal vascular hypothalamic-pituitary system and the neural connection between

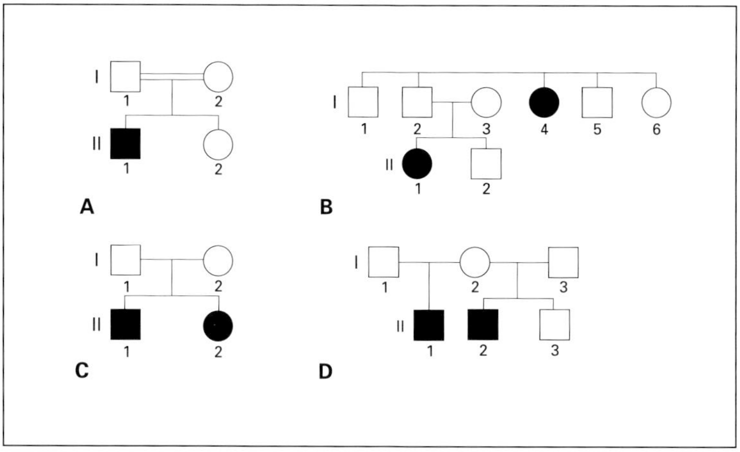

Fig. 4. Familial forms of EPP. ***A*** I1 and I2 are first cousins. II2 has a CPHD involving GH and TSH and an EPP on MRI examination. The consanguinity is in favor of a recessive mode of inheritance of the disease phenotype. ***B*** I4 has a CPHD and II1 has an IGHD. Both patients have an EPP on MRI. The EPP condition is responsible for the anterior pituitary deficiency observed in 2 different members of family B but the type of anterior deficiency (isolated or multiple) is different for the 2 patients. ***C*** II1 and II2 have both an IGHD and an EPP on MRI. There is no known consanguinity. ***D*** II1 and II2 have an IGHD and an EPP. They have the same mother but a different father. An X-linked mode of inheritance is therefore probable.

the median eminence and the posterior pituitary lobe. The 'ectopic bright spot' may then develop as a reparative process after pituitary stalk interruption as observed after surgical transsection in man. In fact, pituitary stalk transsection after postnatal head trauma has been suggested as a cause of acquired GHD in very few cases. However, late growth retardation as mentioned in these reports cannot be regarded as proof of acquired GHD: sufficient blood flow from the hypothalamus to the pituitary in prenatally partially interrupted stalk, may sustain normal GH secretion and growth rate during several postnatal years.

Isolated Changes of Anterior Pituitary Size

MRI of patients with hypopituitarism, most of which presented with IGHD, may show an isolated hypoplasia (fig. 5B) or aplasia (fig. 5A) of the anterior pituitary with a normally located posterior pituitary bright spot. Among these cases, congenital absence of the anterior pituitary is a rare and severe disorder. It can be sporadic or familial, isolated or associated with a spectrum of congenital midline abnormalities similar to that observed in patients with EPP [2, 43–45]. Some of

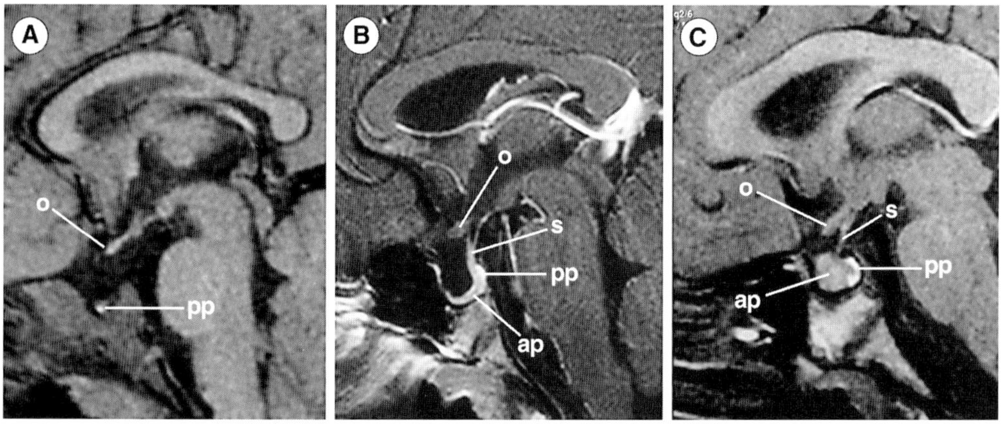

Fig. 5. Various appearances of the anterior pituitary (ap) in nontumoral hypopituitarism with a normally positioned posterior pituitary (pp) (*A* aplasia of the anterior pituitary with a normally positioned posterior pituitary and a nonvisible pituitary stalk, ***B*** hypoplastic anterior pituitary, ***C*** enlargement of the anterior pituitary). o = Optic chiasm, s = pituitary stalk.

these cases have been reported as part of the 'empty sella syndrome', an ill-defined entity more frequently observed in adults. Its clinical expression is quite variable and most of these patients do not show congenital hypopituitarism [46]. The most accepted cause of this 'empty sella' appearance is herniation of the arachnoid space into an enlarged sella turcica.

Another more important issue is the contribution of MRI to the diagnosis of isolated GHD. A recent study of a small number of children with prepubertal GHD indicated that moderate pituitary hypoplasia with intrasellar posterior pituitary does not rule out normalization of GH secretion and of the anterior pituitary volume at adult age [47]. The same outcome was observed in patients with partial GHD and anterior pituitary of normal size at diagnosis. Therefore, in the absence of additional anatomical defects as part of PSIS or of a defined genetic defect (see below), MRI of the pituitary does not significantly contribute to the diagnosis nor to the prognosis of GHD. Hyperplasia of the anterior pituitary has been reported in rare cases of sporadic or familial congenital hypopituitarism. It was associated with normal posterior pituitary and pituitary stalk (fig. 5C). The enlarged anterior pituitary may extend into the cistern and may show cystic areas. It may mislead to the diagnosis of craniopharyngioma. A secondary involution may result in a severely hypoplastic anterior pituitary [5, 32]. This sequence of events recalls the phenomenon of pituitary apoplexy [48]. It may be due to a genetic defect, e.g. mutations in *PROP1*.

Table 2. List of the different candidate genes to be investigated, according to phenotype

Pituitary deficiency	MRI	Associated malformations	Genes
GH	Normal anterior pituitary, posterior pituitary in anatomical position		*GH-N*
GH, probable partial Prl	Hypoplastic anterior pituitary, posterior pituitary in anatomical position		*GHRH-R*
GH, Prl, ± TSH	Normal or hypoplastic anterior pituitary, posterior pituitary in anatomical position		*PIT-1*
GH, Prl, TSH, LH, FSH, ± ACTH	Normal, hypoplastic, or hyperplastic anterior pituitary, posterior pituitary in anatomical position		*PROP-1*
GH, Prl, TSH, LH, FSH, ACTH	Hypoplastic anterior pituitary, posterior pituitary in anatomical position or ectopic	Septo-optic dysplasia	*HESX1*
GH, Prl, TSH, LH, FSH	Hypoplastic or hyperplastic anterior pituitary, posterior pituitary in anatomical position	Stubby neck with rigid cervical spine	*LHX3*

MRI of the Hypothalamic-Pituitary Region in Defined Genetic Defects Causing Hypopituitarism

In recent years, several gene mutations have been shown to cause hypopituitarism. They belong to two groups of genetic abnormalities resulting in growth retardation with GHD: (i) the abnormalities of genes involved in anterior pituitary development and (ii) those involving genes expressed along the somatotropic axis [49].

The nature of the hormonal deficiency (isolated or multiple), as well as the associated phenotypic abnormalities are important in determining which genes should be analyzed. In contrast, the MRI findings of the hypothalamic-pituitary region are unfortunately rarely taken into account when deciding which candidate gene should be investigated. And yet, the anatomy of the hypothalamic-pituitary region may be an additional criteria for selecting the candidate gene defect (table 2).

Genes Involved in Anterior Pituitary Development

So far, abnormalities in four genes, encoding transcription factors (i.e. HESX1, LHX3, PIT-1 and PROP-1) involved in pituitary development, have been described [5, 33]. In the mouse, expression of HESX1 begins before that of LHX3, which in turn is expressed before that of PROP-1 and PIT-1. In humans, mutations of these four different factors have been shown to cause various combinations of anterior pituitary hormone deficiencies. In addition, mutations in *HESX1* may be associated with septo-optic dysplasia [50], whereas the two mutations described so far in *LHX3* are associated with a rigid cervical spine [33]. Patients with mutations in *PIT-1* and *PROP-1* do not display any extrapituitary abnormalities [51]. Patients with HESX1 abnormalities show a hypoplastic anterior pituitary with a posterior pituitary in normal or ectopic position [52a]. It is of interest to note, however, that in mice generated by targeted disruption of *Hesx1*, the anterior pituitary is small and incompletely connected with the posterior pituitary, a finding that is reminiscent of the EPP described in many patients with hypopituitarism [50]. The anterior pituitary is aplastic in mice generated by targeted disruption of *Lhx3* [52b]. In contrast, in the two families described with a *LHX3* mutation [33], the anterior pituitary is either hypoplastic or hyperplastic. In these patients, the posterior pituitary is in a normal anatomical position, as is the case for mice generated by targeted disruption of *Lhx3*. No molecular abnormalities of *LHX3* have been detected in the over 50 patients with EPP and pituitary stalk interruption who were evaluated for this gene [53, 54]. Likewise, the posterior pituitary is always in the anatomical position for patients described thus far with *PIT-1* or *PROP-1* molecular abnormalities, while the anterior pituitary may be either normal or hypoplastic for *PIT-1* and normal, hypoplastic or hyperplastic for *PROP-1* [5]. The naturally occurring mouse models with molecular abnormalities in *Pit-1* [55] and *Prop-1* [56] genes have hypoplastic anterior pituitaries reflecting the lack of somatotrophs, lactotrophs and thyrotrophs (for *Pit-1* and *Prop-1)* associated with the lack of gonadotrophs (for *Prop-1*). The striking enlargement of the anterior pituitary described in some patients with *PROP-1* [32] or *LHX3* [33] mutations has not been described so far in the animal models with abnormalities of these genes. Although the mechanism underlying this phenomenon is still unknown, the existence of the same loss-of-function mutation (delA301,G302) in *PROP-1* associated either with a pituitary hypoplasia [5] or a pituitary enlargement [32] strongly suggests that the nature of the mutation is not the key element in determining the pituitary height. This phenomenon can be observed with abnormalities in at least two different genes encoding transcription factors of the temporospatial cascade involved in anterior pituitary development. One hypothesis is that abnormalities of genes involved in the differentiation of the specialized cells of the anterior pituitary might lead to the growth of undifferentiated tissue by interfering with the control of the hypothalamic feedback regulation.

Genes Expressed along the Somatotropic Axis

To date, molecular abnormalities in only two genes (i.e., *GH-N* encoding the growth hormone and *GHRH-R* encoding the growth hormone-releasing hormone receptor) have been identified in patients with GHD. In one reported patient with a deletion of the entire *GH-N* gene and therefore with a complete absence of growth hormone secretion, the anterior pituitary presented a normal height and the posterior pituitary was in the normal position [57]. A few years after the description of a *GHRH-R* gene mutation in the *little* mouse, a naturally occurring murine strain, two different human mutations were reported [58–61]. In the *little* mouse, pituitary size is normal at birth but progressively becomes hypoplastic during the postnatal period, when GHRH-dependent somatotroph proliferation should normally occur [62]. For one of the described human mutations (E72X), a severely hypoplastic anterior pituitary with a posterior pituitary in the normal anatomical position was found on MRI [59, 63]. However, the volume of the sella turcica appeared normal suggesting an early normal development of both the anterior pituitary and the sella, with a secondary failure of pituitary growth [63]. In humans as in mice, somatotrophs account for approximately half of the anterior pituitary cell mass and the hypoplastic anterior pituitary in humans with a mutation of the *GHRH-R* is consistent with the documented somatotroph depletion of the anterior pituitary in the *little* mouse. A list of the different candidate genes to be investigated, according to phenotype is described in table 2.

Conclusion

Different MRI findings are seen in nontumoral hypopituitarism. Careful imaging of the hypothalamic-pituitary region provides valuable information for the diagnosis of GHD and its possible etiologies. EPP is a common finding in patients with nontumoral hypopituitarism, essentially in CPHD and less frequently in IGHD. It is a marker of permanent GHD and may be the consequence of an early developmental defect. In great majority of cases, a genetic origin has not yet been found. In the recent years, genetic abnormalities have been identified in a small number of patients with severe hypopituitarism, showing a posterior pituitary in the normal anatomical position and an anterior pituitary of variable size. In animal models, many different genes have been found to be involved in the pituitary development, some of which have not yet been identified in humans [64]. Other developmental and/or genetic defects will have to be found to explain the vast majority of pituitary changes observed by MRI.

References

1 Procter AM, Phillips JA 3rd, Cooper DN: The molecular genetics of growth hormone deficiency. Hum Genet 1998;103:255–272.

2 Beregszaszi M et al: Nasal pyriform aperture stenosis and absence of the anterior pituitary gland: Report of two cases. J Pediatr 1996;128:858–861.

3 Rosenfeld RG et al: Diagnostic controversy: The diagnosis of childhood growth hormone deficiency revisited. J Clin Endocrinol Metab 1995;80:1532–1540.

4 Triulzi F et al: Evidence of a congenital midline brain anomaly in pituitary dwarfs: A magnetic resonance imaging study in 101 patients. Pediatrics 1994;93:409–416.

5 Parks JS, Brown MR, Hurley DL, Phelps CJ, Wajnrajch MP: Heritable disorders of pituitary development. J Clin Endocrinol Metab 1999;84:4362–4370.

6 Ikeda H, Suzuki J, Sasano N, Niizuma H: The development and morphogenesis of the human pituitary gland. Anat Embryol 1988;178:327–336.

7 Dasen JS, Rosenfeld MG: Combinatorial codes in signaling and synergy: Lessons from pituitary development. Curr Opin Genet Dev 1999;9:566–574.

8 Hamilton J et al: Familial growth hormone deficiency associated with MRI abnormalities. Am J Med Genet 1998;80:128–132.

9 Argyropoulou M, Perignon F, Brunelle F, Brauner R, Rappaport R: Height of normal pituitary gland as a function of age evaluated by magnetic resonance imaging in children. Pediatr Radiol 1991;21:247–249.

10 Kornreich L, Horev G, Lazar L, Josefsberg Z, Pertzelan A: MR findings in hereditary isolated growth hormone deficiency. AJNR Am J Neuroradiol 1997;18:1743–1747.

11 Takano K, Utsunomiya H, Ono H, Ohfu M, Okazaki M: Normal development of the pituitary gland: Assessment with three-dimensional MR volumetry. AJNR Am J Neuroradiol 1999;20:312–315.

12 Brooks BS, el Gammal T, Allison JD, Hoffman WH: Frequency and variation of the posterior pituitary bright signal on MR images. AJNR Am J Neuroradiol 1989;10:943–948.

13 Abrahams JJ: Ectopia of the posterior pituitary gland or lipoma? AJNR Am J Neuroradiol 1991;12: 579–581.

14 Maghnie M et al: Dynamic MRI in the congenital agenesis of the neural pituitary stalk syndrome: The role of the vascular pituitary stalk in predicting residual anterior pituitary function. Clin Endocrinol (Oxf) 1996;45:281–290.

15 Pellini C et al: Growth hormone deficiency in children: Role of magnetic resonance imaging in assessing aetiopathogenesis and prognosis in idiopathic hypopituitarism. Eur J Pediatr 1990;149: 536–541.

16 Bressani N et al: Evidence of morphological and functional abnormalities in the hypothalamus of growth-hormone-deficient children: A combined magnetic resonance imaging and endocrine study. Horm Res 1990;34:189–192.

17 Cacciari E et al: Endocrine function and morphological findings in patients with disorders of the hypothalamo-pituitary area: A study with magnetic resonance. Arch Dis Child 1990;65:1199–1202.

18 Abrahams JJ, Trefelner E, Boulware SD: Idiopathic growth hormone deficiency: MR findings in 35 patients. AJNR Am J Neuroradiol 1991;12:155–160.

19 Kuroiwa T et al: MR imaging of pituitary dwarfism. AJNR Am J Neuroradiol 1991;12:161–164.

20 Maghnie M et al: Hypothalamic-pituitary dysfunction in growth hormone-deficient patients with pituitary abnormalities. J Clin Endocrinol Metab 1991;73:79–83.

21 Ochi M, Morikawa M, Yoshimoto M, Kinoshita E, Hayashi K: Growth retardation due to idiopathic growth hormone deficiencies: MR findings in 24 patients. Pediatr Radiol 1992;22:477–480.

22 Marwaha R et al: Hypothalamo-pituitary axis by magnetic resonance imaging in isolated growth hormone deficiency patients born by normal delivery. J Clin Endocrinol Metab 1992;74:654–659.

23 Argyropoulou M, Perignon F, Brauner R, Brunelle F: Magnetic resonance imaging in the diagnosis of growth hormone deficiency. J Pediatr 1992;120:886–891.

24 Bozzola M et al: Role of magnetic resonance imaging in the diagnosis and prognosis of growth hormone deficiency. Clin Endocrinol (Oxf)1996;45:21–26.

25 Nagel BH, Palmbach M, Petersen D, Ranke MB: Magnetic resonance images of 91 children with different causes of short stature: Pituitary size reflects growth hormone secretion. Eur J Pediatr 1997;156:758–763.
26 Arrigo T et al: Relationships between neuroradiological and clinical features in apparently idiopathic hypopituitarism. Eur J Endocrinol 1998;139:84–88.
27 Kornreich L et al: MR findings in growth hormone deficiency: Correlation with severity of hypopituitarism. AJNR Am J Neuroradiol 1988;19:1495–1499.
28 Abernethy LJ: Imaging of the pituitary in children with growth disorders. Eur J Radiol 1998;26: 102–108.
29 Hamilton J, Blaser S, Daneman D: MR imaging in idiopathic growth hormone deficiency. AJNR Am J Neuroradiol 1988;19:1609–1615.
30 Arifa N, Leger J, Garel C, Czernichow P, Hassan M: Cerebral anomalies associated with growth hormone insufficiency in children: Major markers for diagnosis? Arch Pediatr 1999;6:14–21.
31 Chen S, Leger J, Garel C, Hassan M, Czernichow P: Growth hormone deficiency with ectopic neurohypophysis: Anatomical variations and relationship between the visibility of the pituitary stalk asserted by magnetic resonance imaging and anterior pituitary function. J Clin Endocrinol Metab 1999;84:2408–2413.
32 Mendonca BB et al: Longitudinal hormonal and pituitary imaging changes in two females with combined pituitary hormone deficiency due to deletion of A301,G302 in the PROP1 gene. J Clin Endocrinol Metab 1999;84:942–945.
33 Netchine I et al: Mutations in LHX3 result in a new syndrome revealed by combined pituitary hormone deficiency. Nat Genet 2000;25:182–186.
34 Fujisawa I et al: Transection of the pituitary stalk: Development of an ectopic posterior lobe assessed with MR imaging. Radiology 1987;165:487–489.
35 Kelly WM et al: Posterior pituitary ectopia: An MR feature of pituitary dwarfism. AJNR Am J Neuroradiol 1988;9:453–460.
36 Simmons GE, Suchnicki JE, Rak KM, Damiano TR: MR imaging of the pituitary stalk: Size, shape, and enhancement pattern. AJR Am J Roentgenol 1992;159:375–377.
37 Genovese E et al: Hypothalamic-pituitary vascularization in pituitary stalk transection syndrome: Is the pituitary stalk really transected? The role of gadolinium-DTPA with spin-echo T1 imaging and turbo-FLASH technique. Pediatr Radiol 1997;27:48–53.
38 di Natale B, Pellini C, Ackermann S, De Angelis R, Chiumello G: Persisting functional connection in growth-hormone-deficient patients with a transected stalk. Horm Res 1994;41:193–196.
39 Ultmann MC, Siegel SF, Hirsch WL, Finegold DN, Foley TP Jr: Pituitary stalk and ectopic hyperintense T1 signal on magnetic resonance imaging. Implications for anterior pituitary dysfunction. Am J Dis Child 1993;147:647–652.
40 Pinto G et al: Pituitary stalk interruption syndrome: A clinical-biological-genetic assessment of its pathogenesis. J Clin Endocrinol Metab 1997;82:3450–3454.
41 Brodsky MC: Optic nerve hypoplasia with posterior pituitary ectopia: Male predominance and nonassociation with breech delivery. Am J Ophthalmol 1999;127:238–239.
42 Siegel SF, Ahdab-Barmada M, Arslanian S, Foley TP Jr: Ectopic posterior pituitary tissue and paracentric inversion of the short arm of chromosome 1 in twins. Eur J Endocrinol 1995;133:87–92.
43 Heinrichs C et al: Fetal hypopituitarism: Perinatal endocrine and morphological studies in two cases. Acta Paediatr 1994;83:448–451.
44 Kato F, Kikuchi K, Miyamoto S, Ohie T, Yamaguchi S: Congenital hypopituitarism with hypoplasia of the anterior pituitary gland and a normal posterior pituitary lobe. Acta Paediatr 1995;84: 1201–1203.
45 Kosaki K, Matsuo N, Tamai S, Miyama S, Momoshima S: Isolated aplasia of the anterior pituitary as a cause of congenital panhypopituitarism. Case report. Horm Res 1991;35:226–228.
46 Ammar A, Al-Sultan A, Al Mulhim F, Al Hassan AY: Empty sella syndrome: Does it exist in children? J Neurosurg 1999;91:960–963.
47 Maghnie M et al: Growth hormone (GH) deficiency (GHD) of childhood onset: Reassessment of GH status and evaluation of the predictive criteria for permanent GHD in young adults. J Clin Endocrinol Metab 1999;84:1324–1328.

48 Pinto G, Zerah M, Trivin C, Brauner R: Pituitary apoplexy in an adolescent with prolactin-secreting adenoma. Horm Res 1998;50:38–41.
49 Lopez-Bermejo I, Buckway CK, Rosenfeld RG: Genetic defects of the growth hormone-insulin-like growth factor axis. Trends Endocrinol Metab 2000;11:39–49.
50 Dattani MT et al: Mutations in the homeobox gene HESX1/Hesx1 associated with septo-optic dysplasia in human and mouse. Nat Genet 1998;19:125–133.
51 Pfäffle RW, Blankenstein O, Wuller S, Kentrup H: Combined pituitary hormone deficiency: Role of Pit-1 and Prop-1. Acta Paediatr Suppl 1999;88:33–41.
52a Thomas PQ, et al: Heterozygous HESX1 mutations associated with isolated congenital pituitary hypoplasia and septo-optic dysplasia. Hum Mol Genet 2001;10:39–45.
52b Sheng HZ, et al: Specification of pituitary cell lineages by the LIM homeobox gene Lhx3. Science 1996;272:1004–1007.
53 Sloop KW, Walvoord EC, Showalter AD, Pescovitz OH, Rhodes SJ: Molecular analysis of LHX3 and PROP-1 in pituitary hormone deficiency patients with posterior pituitary ectopia. J Clin Endocrinol Metab 2000;85:2701–2708.
54 Schmitt S, Biason-Lauber A, Betts D, Schoenle EJ: Genomic structure, chromosomal localization, and expression pattern of the human LIM-homeobox3 (LHX 3) gene. Biochem Biophys Res Commun 2000;274:49–56.
55 Li S et al: Dwarf locus mutants lacking three pituitary cell types result from mutations in the POU-domain gene pit-1. Nature 1990;347:528–533.
56 Sornson MW et al: Pituitary lineage determination by the Prophet of Pit-1 homeodomain factor defective in Ames dwarfism. Nature 1996;384:327–333.
57 Zucchini S, Ambrosetto P, Baroncini C, Cacciari E: Normal pituitary size in two patients with growth hormone gene deletion. J Pediatr Endocrinol Metab 1996;9:545–548.
58 Wajnrajch MP, Gertner JM, Harbison MD, Chua SC Jr, Leibel RL: Nonsense mutation in the human growth hormone-releasing hormone receptor causes growth failure analogous to the little (lit) mouse. Nat Genet 1996;12:88–90.
59 Netchine I et al: Extensive phenotypic analysis of a family with growth hormone (GH) deficiency caused by a mutation in the GH-releasing hormone receptor gene. J Clin Endocrinol Metab 1998; 83:432–436.
60 Maheshwari HG, Silverman BL, Dupuis J, Baumann G: Phenotype and genetic analysis of a syndrome caused by an inactivating mutation in the growth hormone-releasing hormone receptor: Dwarfism of Sindh. J Clin Endocrinol Metab 1998;83:4065–4074.
61 Salvatori R et al: Familial dwarfism due to a novel mutation of the growth hormone-releasing hormone receptor gene. J Clin Endocrinol Metab 1999;84:917–923.
62 Lin SC et al: Molecular basis of the little mouse phenotype and implications for cell type-specific growth. Nature 1993;364:208–213.
63 Murray RA, Maheshwari HG, Russell EJ, Baumann G: Pituitary hypoplasia in patients with a mutation in the growth hormone-releasing hormone receptor gene. AJNR Am J Neuroradiol 2000; 21:685–689.
64 Burrows H, Douglas K, Seasholtz A, Camper S: Genealogy of anterior pituitary gland: Tracing a family tree. Trends Endocrinol Metab 1999;10:343–352.

Irène Netchine, Laboratoire d'Explorations Endocriennes, Hôpital Troussseau, Paris,
et Institut National de la Santé et de la Recherche Médicale (Unité 468)
Hôpital Henri-Mondor, F–94010 Créteil (France)
Tel. +33 1 49 81 28 96, Fax +33 1 48 99 33 45, E-Mail irene.netchine@im3.inserm.fr

Rappaport R, Amselem S (eds): Hypothalamic-Pituitary Development. Genetic and Clinical Aspects.
Endocr Dev. Basel, Karger, 2001, vol 4, pp 109–121

Kallmann Syndrome: Towards Molecular Pathogenesis

Jean-Pierre Hardelin

Unité de Génétique des Déficits Sensoriels, CNRS URA 1968, Institut Pasteur, Paris, France

Gonadotropin-releasing hormone (GnRH) is a key regulator of reproduction and sexual behaviour. During the last decade, embryological studies have clarified the question of the early development of GnRH-synthesizing neurones before the onset of neurosecretion. These studies have revealed the existence of a topographical link between GnRH-synthesizing neurones and the embryonic olfactory system, thereby shedding new light on Kallmann syndrome, a developmental disease characterized by the association of hypogonadotropic hypogonadism and anosmia (or hyposmia). The identification of the gene underlying the X-chromosome-linked form of the disease 10 years ago, has opened the way to molecular pathophysiology.

A Definition of the Syndrome

In 1856, Maestre de San Juan [1], a Spanish anatomist, reported the association of the absence of olfactory structures in the brain and the presence of small testes in an individual. The syndrome was identified as a clinical entity in 1944, by an American medical geneticist, Kallmann, who carried out a study on the occurrence of hypogonadism accompanied by anosmia in three affected families [2]. He showed the co-segregation of the anosmia and the hypogonadism in all the affected individuals and therefore established that this syndrome can be hereditary. In the 1950s, de Morsier [3–5], an anatomist from Switzerland, further documented this syndrome. He described the underdevelopment or absence of the olfactory bulbs and tracts in several male patients with hypogonadism. These olfactory anomalies can be shown by magnetic resonance imaging (MRI) [6–11];

it is noteworthy, however, that normal images have been reported in some male patients affected by Kallmann syndrome [11]. Despite this reservation, the usefulness of MRI in diagnosing the disease in patients too young to undergo meaningful testing of olfaction or of the hypothalamo-pituitary-gonadal axis should be emphasized [8, 12]. De Morsier and Gauthier [13] also found hypothalamic dysplasia and/or hypoplasia, even though no clear anomaly of this brain region has been documented by MRI since that time [6]. The occasional presence of various cerebral or cranial anomalies was also reported [13]. Some years later, the hypogonadism was ascribed to GnRH deficiency [14], which can be estimated in the patients by using several different endocrinal tests (see Seminara et al. [15]). The extent of the hypogonadism and that of the smell deficiency may vary significantly between patients, even between monozygotic twins [16, 17]. However, there is so far no genetic or pathophysiological argument to include isolated (i.e. with normal sense of smell) hypogonadism of hypothalamic origin in the Kallmann syndrome nosological entity. Finally, a variety of other symptoms, both neurological (mirror movements, hearing loss, abnormal visual attention, oculomotor abnormalities, ptosis, cerebellar syndrome, pes cavus, etc.) and non-neurological (renal aplasia, high-arched palate, labial or palatine cleft, dental agenesis, etc.) may be observed [18–26].

The prevalence of the disease has been estimated at 1 out of 10,000 in boys; in girls, the prevalence may be 5–7 times lower [15, 27]. Most cases are sporadic and are diagnosed in adolescence, when an absence of spontaneous puberty is found concomitantly with a deficient sense of smell. The latter should be ascertained by the means of detailed questioning and smell perception tests [28, 29] because it is rarely mentioned spontaneously. On rare occasions, the disease can be diagnosed as early as infancy in the presence of cryptorchidism or a microphallus in boys. In the rare familial forms of the disease, three modes of inheritance have been described: (i) X-chromosome linked (KAL-1, MIM#308700); (ii) autosomal dominant (KAL-2, MIM#147950) and (iii) autosomal recessive (KAL-3, MIM#244200) [30]. In the X-linked form of the disease, carrier females can suffer from hyposmia.

Embryonic Migration of the GnRH-Synthesizing Neurones

During the last decade, numerous embryological studies conducted in mammals [31–34], birds [35–37], amphibians [38, 39] and fish [40] have established the transient existence of a close topographical link between GnRH-synthesizing neurones and the peripheral olfactory system. Indeed, GnRH neurones migrate from the medial part of the olfactory epithelium up to the brain along the olfactory nerve pathway during embryogenesis (fig. 1A). In mammals, they migrate in

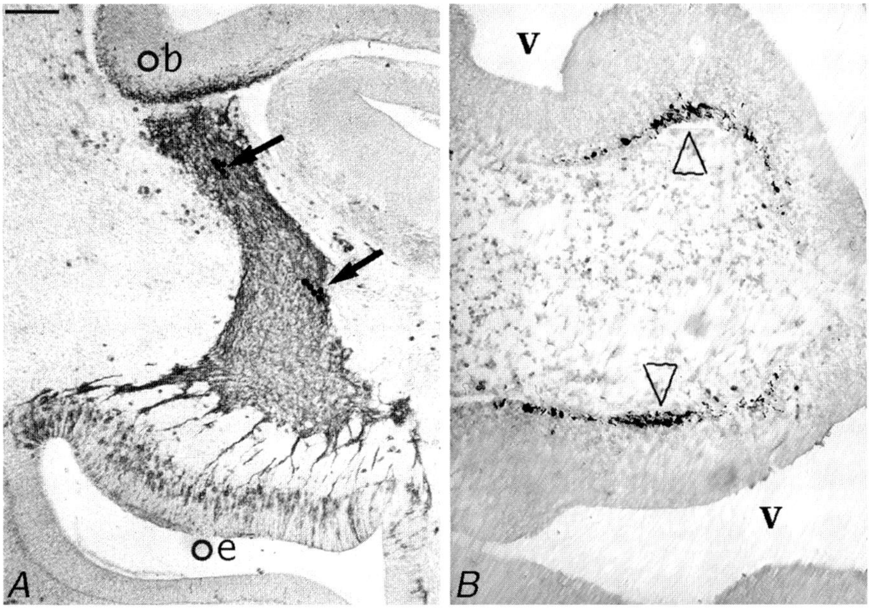

Fig. 1. Embryonic migration of GnRH-synthesizing neurones. ***A*** From the olfactory epithelium to the rostral forebrain: Parasagittal section of the nasal region in a 42-day-old human embryo (low-power photomicrograph). Rostral is on the left. Two clusters of migrating GnRH cells (represented by black dots) can be seen along the broad swath of N-CAM-immunoreactive nerve fibres which extend from the olfactory epithelium (oe) to the developing olfactory bulb (ob). Scale bar = 100 μm. [Adapted from 34, with permission.] ***B*** Intracerebral path: Horizontal section of the primitive cerebral hemispheres in a 42- to 44-day-old embryo (low-power photomicrograph). Rostral is on the left. In each hemisphere, cords of migrating immunoreactive GnRH cells can be seen in the most superficial zone of the medial neural wall (arrowhead). v = Lateral ventricles. Scale bar = 150 μm. [Reproduced from 71, with permission.]

close association with fibres of the vomeronasal and terminal nerves [41–44]. On reaching the base of the telencephalon, they penetrate into the brain, just caudal to the olfactory bulb anlage, then make their way superficially along the medial wall of the cerebral hemisphere [44, 45] (fig. 1B), to the hypothalamic region where they finally settle [46, 47]. Although GnRH neurones enter the brain with the central processes of the terminal nerves, there is still some uncertainty as to whether these nerve fibres guide the GnRH cells in their intracerebral navigation. In man, migration of the GnRH-synthesizing neurones begins during the sixth embryonic week [34]. This period is also when axon terminals of the olfactory receptor neurones first come into contact with the rostral pole of the forebrain

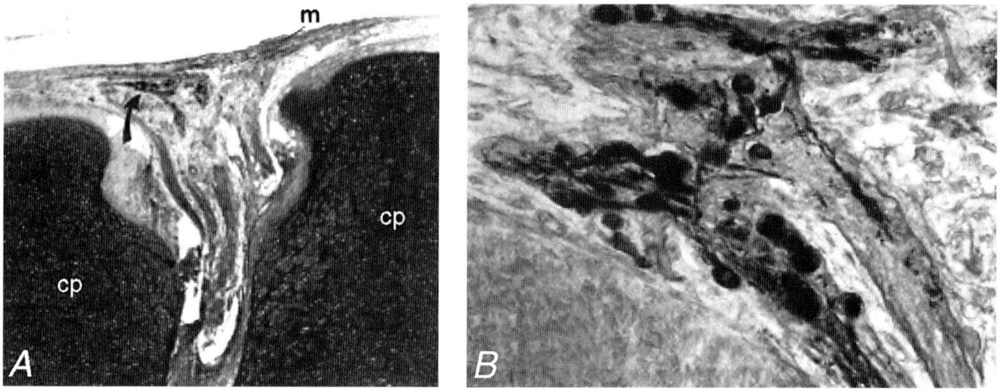

Fig. 2. Failed migration of GnRH cells in X-linked Kallmann syndrome. ***A*** Low magnification of a sagittal section through the upper nasal region of a 19-week-old human foetus with X-linked Kallmann syndrome. A GnRH-immunoreactive fascicule of the terminal nerve emerges from the nasal cavity through a perforation in the cribriform plate (cp) of the ethmoid bone. A cluster of GnRH cells (arrow) is seen between the dorsal surface of cribriform plate and the meninges (m). No GnRH cells were found in the brain of this foetus and both olfactory bulbs were absent. ***B*** At higher magnification, thick clusters of immunoreactive GnRH cells are seen in ganglia of the terminal nerve. [Reproduced from 44, with permission.]

[48]; this initial contact occurs just before the emergence of the olfactory bulbs [48–50] and is a condition of their differentiation [41, 51].

Disrupted Migration of the GnRH Neurones in a Foetus Affected by Kallmann Syndrome

The histopathological examination of a 19-week-old male foetus carrying a deletion of the *KAL-1* gene (responsible for the X-linked form of the syndrome) revealed both an absence of the olfactory bulbs and an abnormal localisation of the GnRH-synthesizing neurones. In this foetus, GnRH neurones had left the olfactory epithelium, but had accumulated outside the brain in the upper nasal region, along with the axon terminals of the olfactory, vomeronasal and terminal nerves [52] (fig. 2). Although the embryonic stage and precise site of the primary defect cannot be deduced from these limited observations, they seemed to point towards the involvement of *KAL-1* in the final migration stretch travelled by these cells and axonal fibres. On the other hand, a crucial role of the gene in the initial part of the migrating path seemed to be ruled out.

Identification of the *KAL-1* Gene, and Clinical Correlations

KAL-1 has been isolated through positional cloning approaches [53, 54], and validated as the actual gene responsible for the X-linked form of Kallmann syndrome by the finding of mutations in affected families [55–57]. A variety of nonsense, missense and frameshift point mutations as well as large gene deletions have been reported since then [58]. Moreover, clinical analysis of individuals carrying *KAL-1* mutations has made it possible to ascribe synkineses (mirror movements), renal aplasia, high-arched palate, deafness and pes cavus, to the X-linked form of the disease [57]. Mirror movements and renal aplasia are often encountered; they affect three quarters, and a third of patients, respectively [59–61]. Mirror movements seem to be specific to the X-linked form of the disease; abdominal echography, ideally performed in all patients affected by Kallmann syndrome, should soon resolve the question of whether renal aplasia is also characteristic of this genetic form.

It is interesting to note that the stronger prevalence of Kallmann syndrome in males does not seem to be due to the frequent involvement of *KAL-1* as may have been believed a priori. Indeed, molecular analysis of this gene in the far more numerous sporadic cases of the disease only very rarely reveals the presence of a mutated allele. Therefore, half a century after the first description of Kallmann syndrome as a hereditary disease [2], the genetics of the syndrome remain poorly understood. In particular, the contribution of the autosomal forms of the disease (KAL-2 and KAL-3 [30]) to the sporadic cases and the role of epigenetic factors need to be clarified.

Structure of the Encoded Protein, Anosmin-1: Conservation among Animal Species

The *KAL-1*-deduced amino acid sequence (680 amino acids) predicts an extracellular glycoprotein of compound modular structure: (i) a cysteine-rich amino-terminal region, including a whey acidic protein-like four-disulfide core motif which is present in several secreted proteins having inhibitory activity for serine proteases [62–64]; (ii) four repetitions of the fibronectin-like type III motif [65], with high homology to similar motifs of certain axonal adhesion molecules, and (iii) a short carboxy-terminal region rich in basic amino acids (fig. 3). The protein has been named anosmin-1 in reference to the defective sense of smell characterizing the disease.

KAL-1 orthologues have now been identified in several vertebrates (chicken, zebrafish) [66, 67] and invertebrates *(Drosophila, Caenorhabditis elegans)* [68]. The predicted proteins have the same aforementioned arrangement of a whey

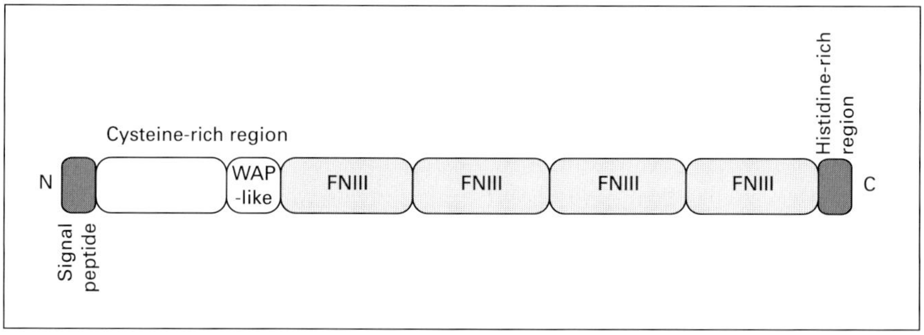

Fig. 3. Predicted modular structure of anosmin-1. WAP-like and FNIII refer to the whey acidic protein-like domain and fibronectin-like type III repeats, respectively.

acidic protein-like motif and four fibronectin-like type III repeats, suggesting that their roles may have been conserved during evolution. It is noteworthy that the orthologous genes have so far escaped detection in rodents, most likely because of a highly divergent nucleotide sequence [69]. This unusual situation has prevented the generation of a mouse model of the disease by gene knockout.

Anosmin-1 Is a Locally Restricted Component of Extracellular Matrices during the Organogenesis Period

Specific polyclonal and monoclonal antibodies have been generated against the entire protein (100 kDa) purified from a mammalian cell line which had been transfected by *KAL-1* cDNA [70]. The expression of *KAL-1* in the human embryo was analysed during the organogenesis period by both in situ hybridization and immunohistofluorescence experiments [71]. Between 4 and 10 weeks of embryonic development, the transcript and the protein were detected in various structures of the digestive, respiratory, urogenital, cardiovascular, musculo-skeletal and integumentary systems, in some placodal derivatives and in the central nervous system. During this period of development, anosmin-1 is detected as a regionally restricted component of basement membranes and interstitial matrices. For example, figure 4 illustrates the presence of the protein in the basement membranes of the branches of the ureteric bud. Later in embryonic life, both in the chicken and in man, *KAL-1* expression apparently becomes restricted to definite neuronal populations [70, 72–74].

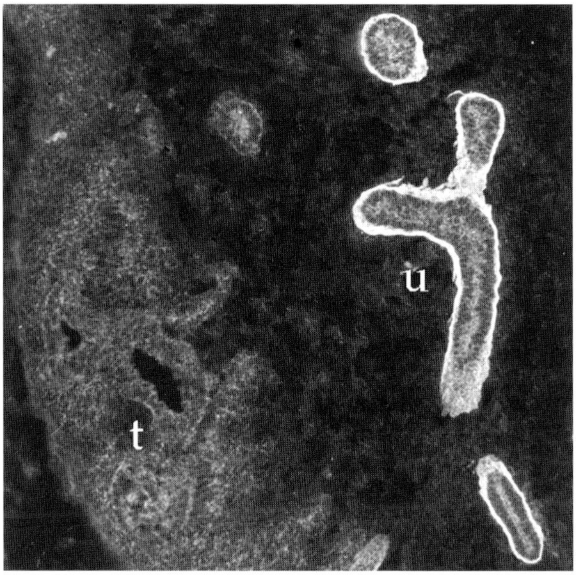

Fig. 4. Anosmin-1 in the embryonic kidney. Immunohistofluorescence in the early metanephros (week 7). Anosmin-1 is detected in the basement membranes of the proximal branches of the ureteric bud (u). In contrast, no immunolabeling is observed in the cortical zone at this stage, either in the distal ureteric branches or in the differentiating secretory tubules (t). This expression pattern suggests that renal aplasia which is present in about one third of the males affected by X-linked Kallmann syndrome, results from primary failure of the collecting duct system. [Reproduced from 71, with permission.]

Anosmin-1 in the Early Olfactory System: Pathophysiological Implications

In the olfactory system, anosmin-1 was detected as early as the fifth embryonic week in the presumptive olfactory bulb region of the telencephalon, where the protein is thus present at the time of the first contact with the olfactory axons in the 6th week (fig. 5). The protein could still be detected in the primitive olfactory bulbs. On the other hand, anosmin-1 was not detected in either the olfactory epithelium or the initial stretch travelled by the GnRH cells and the axons of the olfactory receptor neurones on their route to the brain [71]. Therefore, we propose that anosmin-1 is required for the intracerebral elongation and/or the targeting of the early olfactory axons. As a corollary, we suggest that the genetic defect underlying X-linked Kallmann syndrome disrupts the terminal navigation of these axons. According to this hypothesis, failed differentiation of the olfactory

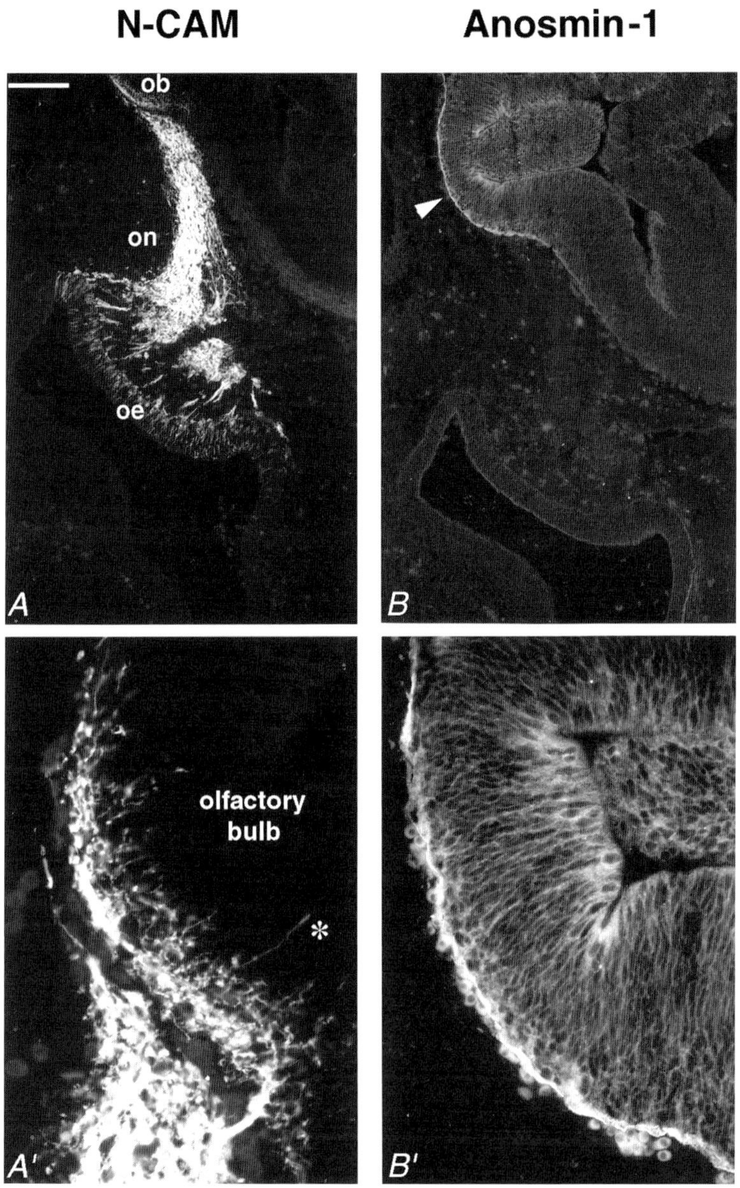

Fig. 5. Anosmin-1 in the early olfactory system. Immunofluorescence on parasagittal sections of the frontonasal region in a 6-week-old embryo (*A*, *B*), and detail of the olfactory bulb anlage (*A′*, *B′*). In *A*, an antibody to N-CAM detects the olfactory nerve (on) fibres, which connect the olfactory epithelium (oe) to the olfactory bulb (ob). Some pioneer olfactory axons can be seen entering the neural wall (asterisk in *A′*). In *B*, an antibody to anosmin-1 detects the protein in the olfactory bulb anlage (arrowhead); anosmin-1 is located both in the

bulbs in the disease would result from the failure of primary contacts between the olfactory axons and the rostral forebrain. However, the possibility that anosmin-1 plays a direct role in the initial steps of the olfactory bulb differentiation should also be considered. Anosmin-1 was also detected at 6 weeks in the medial walls of the primitive cerebral hemispheres, i.e. on the path of the GnRH-synthesizing neurones and central roots of terminal nerves (fig. 1B). Therefore, since these nerves appear to guide the GnRH cells, at least up to their entry into the brain [42–44], the blocked migration of the GnRH cells which was observed in the Kallmann affected foetus (fig. 2) may have resulted from the disruption of the central roots of the terminal nerves as they were penetrating the forebrain. The pattern of anosmin-1 expression in the olfactory system at later developmental stages remains to be determined.

Perspectives on Functional Approaches

Two properties of anosmin-1 have been found in vitro: it is an adhesive substrate for various cell types and it is permissive for neurite growth in diverse neuronal populations [70, 75]. However, the physiological meaning of these findings is still unclear. The putative serine protease inhibitory activity of the protein, suggested by its whey acidic protein-like motif, still awaits experimental demonstration. In addition, the possibility that anosmin-1 undergoes a proteolytic cleavage [70, 76], which would be required for its role in vivo, is under investigation. In any case, deciphering the part played by anosmin-1 in the development of the olfactory system and the migration of GnRH-synthesizing neurones eventually relies on the analysis of an animal model of the disease. However, because the murine *Kal1* gene has so far escaped detection, a targeted inactivation of the gene in this species cannot be undertaken for the moment. The zebrafish may represent an interesting alternative. In this species, two orthologues of *KAL-1* have been identified [67], and gene inactivation strategies based on antisense oligonucleotide injection in the early embryo can be considered [77]. Finally, the gene(s) responsible for the other genetic forms of the disease (KAL-2, KAL-3) remain to be identified, and their roles in the development of the GnRH neurosecretion system remain to be elucidated.

interstitial matrix and in the basement membrane (***B′***). In contrast, no immunoreactivity is observed in either the olfactory epithelium or the nasofrontal mesenchyme. Scale bar: 200 µm in ***A***, ***B*** and 50 µm in ***A′***, ***B′***. [Reproduced from 71, with permission.]

References

1 Maestre de San Juan A: Falta total de los nervios olfactorios con anosmia en un individuo en quien existia una atrofia congenita de los testiculos y miembro viril. Siglo Medico 1856;131:211.

2 Kallmann FJ, Schoenfeld WA, Barrera SE: The genetic aspects of primary eunuchoidism. Am J Ment Defic 1944;48:203–236.

3 De Morsier G: Etudes sur les dysraphies crânio-encéphaliques. 1. Agénésie des lobes olfactifs (télencephaloschizis latéral) et des commissures calleuse et antérieure (télencephaloschizis médian). La dysplasie olfacto-génitale. Schweiz Arch Neurol Psychiat 1955;74:309–361.

4 De Morsier G: Median cranioencephalic dysraphias and olfactogenital dysplasia. World Neurol 1962;3:485.

5 De Morsier G, Gauthier G: La dysplasie olfacto-génitale. Pathol Biol 1963;11:1267–1272.

6 Klingmüller D, Duwes W, Krahe T, Brecht G, Schweikert HU: Magnetic resonance imaging of the brain in patients with anosmia and hypothalamic hypogonadism (Kallmann's syndrome). J Clin Endocrinol Metab 1987;65:581–584.

7 Knorr JR, Ragland RL, Brown RS, Gelber N: Kallmann syndrome: MR findings. Am J Neuroradiol 1993;14:845–851.

8 Truwitt CL, Barkovich AJ, Grumbach MM, Martini JJ: MR imaging of Kallmann syndrome, a genetic disorder of neuronal migration affecting the olfactory and genital systems. Am J Neuroradiol 1993;14:827–838.

9 Yousem DM, Turner WJ, Li C, Snyder PJ, Doty RL: Kallmann syndrome: MR evaluation of olfactory system. Am J Neuroradiol 1993;14:839–843.

10 Vogl T, Stemmler J, Heye B, Schopohl J, Danek A, Bergman C, Balzer J, Felix R: Kallmann syndrome versus idiopathic hypogonadotropic hypogonadism at MR imaging. Radiology 1994;191: 53–57.

11 Quinton R, Duke VM, de Zoysa PA, Platts AD, Valentine AR, Kendall BA: The neuroradiology of Kallmann's syndrome: A genotypic and phenotypic analysis. J Clin Endocrinol Metab 1996;81: 3010–3017.

12 Birnbacher R, Wandl-Vergesslich K, Frisch H: Diagnosis of X-recessive Kallmann syndrome in early infancy. Evidence of hypoplastic rhinencephalon. Eur J Pediatr 1994;153:245–247.

13 Gauthier G: La dysplasie olfacto-génitale (agénésie des lobes olfactifs avec absence de développement gonadique à la puberté. Acta Neurov (Wien) 1960;21:345–394.

14 Naftolin F, Harris GW, Bobrow M: Effect of purified luteinizing hormone releasing factor on normal and hypogonadotropic anosmic men. Nature 1971;232:496–497.

15 Seminara S, Hayes F, Crowley W: Gonadotropin-releasing hormone deficiency in the human (idiopathic hypogonadotropic hypogonadism and Kallmann's syndrome): Pathophysiological and genetic considerations. Endocr Rev 1998;19:521–539.

16 Hermanussen M, Sippell WG: Heterogeneity of Kallmann's syndrome. Clin Genet 1985;28:106–111.

17 Hipkin L, Casson I, Davies JC: Identical twins discordant for Kallmann's syndrome. J Med Genet 1990;27:198–199.

18 Wegenke JD, Uehling DT, Wear JBJ, Gordon ES, Bargman JG, Deacon JSR, Herrmann JPR, Opitz JM: Familial Kallmann syndrome with unilateral renal aplasia. Clin Genet 1975;7:368–381.

19 Conrad B, Kriebel J, Hetzel WD: Hereditary bimanual synkinesis combined with hypogonadotropic hypogonadism and anosmia in four brothers. J Neurol 1978;218:263–274.

20 Lieblich JM, Rogol AD, White BJ, Rosen SW: Syndrome of anosmia with hypogonadotropic hypogonadism (Kallmann syndrome). Am J Med 1982;73:506–519.

21 White BJ, Rogol AD, Brown KS, Lieblich JM, Rosen SW: The syndrome of anosmia with hypogonadotropic hypogonadism: A genetic study of 18 new families and a review. Am J Med Genet 1983; 15:417–435.

22 Schwankhaus JD, Currie J, Jaffe MJ, Rose SR, Sherins RJ: Neurologic findings in men with isolated hypogonadotropic hypogonadism. Neurology 1989;39:223–226.

23 Kertzman C, Robinson DL, Sherins RJ, Schwankhaus JD, McClurkin JW: Abnormalities in visual spatial attention in men with mirror movements associated with isolated hypogonadotropic hypogonadism. Neurology 1990;40:1057–1063.

24 Hill J, Elliott C, Colquhoun I: Audiological, vestibular and radiological abnormalities in Kallmann's syndrome. J Laryngol Otol 1992;106:530–534.
25 De Zegher F, Lagae L, Declerck D, Vinckier F: Kallmann syndrome and delayed puberty associated with agenesis of lateral maxillary incisors. J Craniofac Genet Dev Biol 1995;15:87–89.
26 Molsted K, Kjær I, Giwercman A, Versterhauge S, Skakkebaek N: Craniofacial morphology in patients with Kallmann's syndrome with and without cleft lip and palate. Cleft Palate Craniofac J 1997;34:417–424.
27 Jones J, Kemmann E: Olfacto-genital dysplasia in the female. Obstet Gynecol Annu 1976;5:443–445.
28 Rosen S, Gann P, Rogol A: Congenital anosmia: Detection thresholds for seven odorant classes in hypogonadal and eugonadal patients. Ann Otol Rhinol Laryngol 1979;88:288.
29 Doty R, Shaman P, Dann M: Development of the University of Pennsylvania Smell Identification Test, a standardized microencapsulated test of olfactory function. Physiol Behav 1984;32:489–502.
30 McKusick VA: Mendelian Inheritance in Man. A Catalog of Human Genes and Genetic Disorders. Baltimore, Johns Hopkins University Press, 1998.
31 Schwanzel-Fukuda M, Pfaff DW: Origin of luteinizing hormone-releasing hormone neurons. Nature 1989;338:161–164.
32 Wray S, Grant P, Gainer H: Evidence that cells expressing luteinizing hormone-releasing hormone mRNA in the mouse are derived from progenitor cells in the olfactory placode. Proc Natl Acad Sci USA 1989;86:8132–8136.
33 Ronnekleiv OK, Resko JA: Ontogeny of gonadotropin-releasing hormone-containing neurons in early fetal development of rhesus macaques. Endocrinology 1990;126:498–511.
34 Schwanzel-Fukuda M, Crossin KL, Pfaff DW, Bouloux PMG, Hardelin JP, Petit C: Migration of luteinizing hormone-releasing hormone neurons in early human embryos. J Comp Neurol 1996;366:547–557.
35 Murakami S, Seki T, Wakabayashi K, Arai Y: The ontogeny of luteinizing hormone-releasing hormone (LHRH) producing neurons in the chick embryo: Possible evidence for migrating LHRH neurons from the olfactory epithelium expressing a highly polysialylated neural cell adhesion molecule. Neurosci Res 1991;12:421–431.
36 Norgren RB, Lehman MN: Neurons that migrate from the olfactory epithelium in the chick express luteinizing hormone-releasing hormone. Endocrinology 1991;128:1676–1678.
37 Murakami S, Arai Y: Direct evidence for the migration of LHRH neurons from the nasal region to the forebrain in the chick embryo: A carbocyanine dye analysis. Neurosci Res 1994;19:331–338.
38 Muske L, Moore FL: Ontogeny of gonadotropin-releasing hormone neuronal systems in amphibians. Brain Res 1990;534:177–187.
39 Murakami S, Kikuyama S, Arai Y: The origin of the luteinizing hormone-releasing hormone neurons in newts *(Cynops pyrrhogaster)*: The effect of olfactory placode ablation. Cell Tissue Res 1992;269:21–27.
40 Chiba A, Oka S, Honma Y: Ontogenetic development of gonadotropin-releasing hormone-like immunoreactive neurons in the brain of the chum salmon, *Oncorhynchus keta*. Neurosci Lett 1994;178:51–54.
41 Pearson AA: The development of the nervus terminalis in man. J Comp Neurol 1941;75:39–66.
42 Schwanzel-Fukuda M, Abraham S, Crossin KL, Edelman GM, Pfaff DW: Immunocytochemical demonstration of neural cell adhesion molecule along the migration route of luteinizing hormone-releasing hormone neurons in mice. J Comp Neurol 1992;321:1–18.
43 Livne I, Gibson MJ, Silverman AJ: Biochemical differentiation and intercellular interactions of migratory gonadotropin-releasing hormone cells in the mouse. Dev Biol 1993;159:643–656.
44 Schwanzel-Fukuda M: The origin and migration of LHRH neurons in mammals: A comparison between species including humans; in Parhar I, Sakuma Y (eds): GnRH Neurons: Gene to Behavior. Tokyo, Brain Shuppan, 1997, pp 221–242.
45 Norgren RB, Brackenbury R: Cell adhesion molecules and the migration of LHRH neurons during development. Dev Biol 1993;160:377–387.
46 Paulin C, Dubois M, Barry J, Dubois P: Immunofluorescence study of LH-RH producing cells in the human fetal hypothalamus. Cell Tissue Res 1977;182:341–345.

47 Grumbach M, Kaplan S: The neuroendocrinology of human puberty: An ontogenetic perspective; in Grumbach M, Sizonenko P (eds): Control of the Onset of Puberty. Baltimore, William & Wilkins, 1990, pp 1–68.
48 Bossy J: Development of olfactory and related structures in staged human embryos. Anat Embryol (Berl) 1980;161:225–236.
49 Muller F, O'Rahilly R: The human brain at stage 17, including the appearance of the future olfactory bulb and the first amygdaloid nuclei. Anat Embryol 1989;180:353–369.
50 Gong Q, Shipley MT: Evidence that pioneer olfactory axons regulate telencephalon cell cycle kinetics to induce the formation of the olfactory bulb. Neuron 1995;14:91–101.
51 Stout RP, Graziadei PPC: Influence of the olfactory placode on the development of the brain in *Xenopus laevis* (Daudin). I. Axonal growth and connections of the transplanted placode. Neuroscience 1980;5:2175–2186.
52 Schwanzel-Fukuda M, Bick D, Pfaff DW: Luteinizing hormone-releasing hormone-expressing cells do not migrate normally in an inherited hypogonadal (Kallmann) syndrome. Mol Brain Res 1989;6: 311–326.
53 Legouis R, Hardelin JP, Levilliers J, Claverie JM, Compain S, Wunderle V, Millasseau P, Le Paslier D, Cohen D, Caterina D, Bougueleret L, Delemarre-Van de Waal H, Lutfalla G, Weissenbach J, Petit C: The candidate gene for the X-linked Kallmann syndrome encodes a protein related to adhesion molecules. Cell 1991;67:423–435.
54 Franco B, Guioli S, Pragliola A, Incerti B, Bardoni B, Tonlorenzi R, Carrozzo R, Maestrini E, Pieretti M, Taillon-Miler P, Brown CJ, Willard HF, Lawrence C, Persico MG, Camerino G, Ballabio A: A gene deleted in Kallmann's syndrome shares homology with neural cell adhesion and axonal path-finding molecules. Nature 1991;353:529–536.
55 Bick D, Franco B, Sherins RJ, Heye B, Pike L, Crawford J, Maddalena A, Incerti B, Pragliola A, Meitinger T, Ballabio A: Intragenic deletion of the *KALIG-1* gene in Kallmann's syndrome. N Engl J Med 1992;326:1752–1755.
56 Hardelin JP, Levilliers J, del Castillo I, Cohen-Salmon M, Legouis R, Blanchard S, Compain S, Bouloux P, Kirk J, Moraine C, Chaussain JL, Weissenbach J, Petit C: X chromosome-linked Kallmann syndrome: Stop mutations validate the candidate gene. Proc Natl Acad Sci USA 1992;89: 8190–8194.
57 Hardelin JP, Levilliers J, Blanchard S, Carel JC, Leutenegger M, Pinard-Bertelletto JP, Bouloux P, Petit C: Heterogeneity in the mutations responsible for X chromosome-linked Kallmann syndrome. Hum Mol Genet 1993;2:373–377.
58 Ballabio A, Rugarli E: Kallmann syndrome; in Scriver CR, Beaudet AL, Sly WS, Valle D (eds): The Metabolic and Molecular Bases of Inherited Disease. Montreal, McGraw-Hill, 2001, chapt 225, pp 5729–5739.
59 Mayston MJ, Harrison LM, Quinton R, Stephens JA, Krams M, Bouloux PM: Mirror movements in X-linked Kallmann's syndrome. I. A neurophysiological study. Brain 1997;120:1199–1216.
60 Leinsinger GL, Heiss DT, Jassoy AG, Pfluger T, Hahn K, Danek A: Persistent mirror movements: Functional MR imaging of the hand motor cortex. Radiology 1997;203:545–552.
61 Kirk J, Grant D, Besser G, Shalet S, Smith C, White M, Edwardes O, Bouloux PM: Unilateral renal aplasia in X-linked Kallmann's syndrome. Clin Genet 1994;46:260–262.
62 Seemüller U, Arnhold M, Fritz H, Wiedenmann K, Machleidt W, Heinzel R, Appelhans H, Gassen HG, Lottspeich F: The acid-stable proteinase inhibitor of human mucous secretions (HUSI-I, antileukoprotease). FEBS Lett 1986;199:43–48.
63 Stetler G, Brewer MT, Thompson RC: Isolation and sequence of a human gene encoding a potent inhibitor of leucocyte proteases. Nucl Acids Res 1986;14:7883–7896.
64 Wiedow O, Schroder JM, Gregory H, Young JA, Christophers E: Elafin: an elastase-specific inhibitor of human skin. Purification, characterization and complete amino acid sequence. J Biol Chem 1990;265:14791–14795.
65 Del Castillo I, Cohen-Salmon M, Blanchard S, Lutfalla G, Petit C: Structure of the X-linked Kallmann syndrome gene and its homologous pseudogene on the Y chromosome. Nat Genet 1992;2: 305–310.

66 Legouis R, Cohen-Salmon M, del Castillo I, Levilliers J, Capy L, Mornon JP, Petit C: Characterization of the chicken and quail homologues of the human gene responsible for the X-linked Kallmann syndrome. Genomics 1993;17:516–518.
67 Ardouin O, Legouis R, Fasano L, David-Watine B, Korn H, Hardelin JP, Petit C: Characterization of the two zebrafish orthologues of the *KAL-1* gene underlying X chromosome-linked Kallmann syndrome. Mech Dev 2000;90:89–94.
68 Rubin G, Yandell M, Wortman J: Comparative genomics of the eukaryotes. Science 2000;287: 2204–2215.
69 Salido EC, Li XM, Yen PH, Martin N, Mohandas TK, Shapiro LJ: Cloning and expression of the mouse pseudoautosomal steroid sulphatase gene (Sts). Nat Genet 1996;13:83–86.
70 Soussi-Yanicostas N, Hardelin JP, Arroyo-Jimenez M, Ardouin O, Legouis R, Levilliers J, Traincard F, Betton JM, Cabanié L, Petit C: Initial characterization of anosmin-1, a putative extracellular matrix protein synthesized by definite neuronal cell populations in the central nervous system. J Cell Sci 1996;109:1749–1757.
71 Hardelin JP, Julliard AK, Moniot B, Soussi-Yanicostas N, Verney C, Schwanzel-Fukuda M, Ayer-Le Lièvre C, Petit C: Anosmin-1 is a regionally restricted component of basement membranes and interstitial matrices during organogenesis: Implications for the developmental anomalies of X chromosome-linked Kallmann syndrome. Dev Dyn 1999;215:26–44.
72 Legouis R, Ayer-Le Lièvre C, Leibovici M, Lapointe F, Petit C: Expression of the *KAL* gene in multiple neuronal sites during chicken development. Proc Natl Acad Sci USA 1993;90:2461–2465.
73 Rugarli EI, Lutz B, Kuratani SC, Wawersik S, Borsani G, Ballabio A, Eichele G: Expression pattern of the Kallmann syndrome gene in the olfactory system suggests a role in neuronal targeting. Nat Genet 1993;4:19–26.
74 Lutz B, Kuratani S, Rugarli EI, Wawersik S, Wong C, Bieber FR, Ballabio A, Eichele G: Expression of the Kallmann syndrome gene in human fetal brain and in the manipulated chick embryo. Hum Mol Genet 1994;3:1717–1723.
75 Soussi-Yanicostas N, Faivre-Sarrailh C, Hardelin JP, Levilliers J, Rougon G, Petit C: Anosmin-1 underlying the X chromosome-linked Kallmann syndrome is an adhesive molecule that can modulate neurite growth in a cell-type specific manner. J Cell Sci 1998;111:2953–2965.
76 Rugarli EI, Ghezzi C, Valsecchi V, Ballabio A: The Kallmann syndrome gene product expressed in COS cells is cleaved on the cell surface to yield a diffusible component. Hum Mol Genet 1996;5: 1109–1115.
77 Hammerschmidt M, Bladder P, Strähle U: Strategies to perturb zebrafish development; in The Zebrafish: Biology. New York, Academic Press, 1999, vol 59, pp 87–116.

Jean-Pierre Hardelin, MD, PhD, Unité de Génétique des Déficits Sensoriels, CNRS URA 1968
Institut Pasteur, 25, rue du Dr Roux, F–75724 Paris Cedex 15 (France)
Tel. +33 1 45 68 88 50, Fax +33 1 45 67 69 78, E-Mail hardelin@pasteur.fr

Rappaport R, Amselem S (eds): Hypothalamic-Pituitary Development. Genetic and Clinical Aspects. Endocr Dev. Basel, Karger, 2001, vol 4, pp 122–139

Genetic Basis of Hypothalamic-Pituitary Hypogonadism

François P. Pralong

Division of Endocrinology, Diabetology and Metabolism, Department of Medicine, University Hospital, Lausanne, Switzerland

Introduction

The activity of the reproductive axis in humans undergoes important dynamic changes during life. In the fetal and neonatal periods until infancy, gonadotropins and gonadal steroids are secreted at levels similar to those seen in early puberty. This secretion then recedes during infancy until the beginning of puberty, when the pituitary synthesis and release of both gonadotropins increases again. This results in the stimulation of gonadal steroid secretion, the appearance and development of secondary sexual characteristics, and eventually in the acquisition of the full fertility potential of normal adults. The activity of the gonadotropic axis then persists throughout adulthood. It is usually very regular in men, and depends upon the stage of the menstrual cycle in women. The delicate sequence of events characterizing normal reproductive activity is critically dependent upon the activity of the hypothalamic gonadotropin-releasing hormone (GnRH): a pulsatile secretion of GnRH into vessels of the hypothalamo-pituitary portal system is mandatory to induce the synthesis and release of the two gonadotropins, luteinizing hormone (LH) and follicle-stimulating hormone (FSH), by gonadotropes of the anterior pituitary gland [1]. The secretion of LH and FSH into the peripheral blood is therefore also pulsatile [2], and LH can be used as a surrogate marker of the pulsatile hypothalamic GnRH secretion [3] in clinical investigation [4]. According to a 'classical' physiological view, LH secreted in the periphery will stimulate gonadal steroid biosynthesis and secretion, whereas FSH regulates gametogenesis in both men and women. As will be detailed below, new insight into the physiological regulation of the gonadotropic axis in humans has been gained in recent years by the careful phenotyping of individuals bearing inactivating mutations of genes acting at different levels in that axis [5–7].

Isolated GnRH Deficiency

Failure of the hypothalamus to secrete GnRH, or absence of pituitary responsiveness to hypothalamic GnRH secretion, results in the clinical syndrome of hypogonadotropic hypogonadism, also referred to as isolated GnRH deficiency [7]. The association of isolated GnRH deficiency and anosmia represents Kallmann's syndrome, whereas patients with a normal sense of smell are usually referred to as having idiopathic hypogonadotropic hypogonadism (IHH). Maestre de San Juan [8], in the 19th century, was the first to describe the association between isolated GnRH deficiency and anosmia, but the merit of Kallmann and Schoenfeld [9] was to clearly ascribe a genetic basis to this association in 1944. Ten years later, de Morsier [10] described the association of isolated GnRH deficiency and anosmia with an agenesis of the olfactory bulbs. This feature can be ascertained with modern imaging techniques [11–13]. It remains a hallmark of Kallmann's syndrome, although up to 25% of Kallmann's syndrome patients have a normal olfactory bulb on magnetic resonance imaging (MRI) studies [14].

The true incidence of isolated GnRH deficiency is very difficult to estimate. In a population of 600,000 Sardinian males subjected to screening for military service over a period of 30 years, 265 were rejected because of hypogonadism [15]. Out of these 265 men, 7 suffered from Kallmann's syndrome, thus providing an estimated incidence of 1 in 86,000 in this particular population. However, the results of a similar screening in a French population were markedly divergent, suggesting an incidence of 1 in 10,000 [16]. Such discrepancies probably illustrate the difficulty to calculate the real prevalence of rare diseases, but they could also partially result from the heterogeneous phenotype of the syndrome [17–21].

One reason accounting for this clinical heterogeneity has become widely recognized over recent years: the phenotypes associated with single-gene mutations of the gonadotropic axis can be variable. Male patients with complete impuberism and undescended testes or female patients with primary amenorrhea and no breast development lie at one extremity of the spectrum, whereas subjects with normal secondary sexual development and subtle defects in gametogenesis represent the other end of phenotypic expression. The study of these patients demonstrates that very good genotype/phenotype correlation can occur for some of these mutations. As will be reviewed below, they have provided valuable information about structure-function relationships of the affected genes and proteins, sometimes allowing to delineate their physiological roles in humans with greater precision.

Clinical Presentation

The diagnosis of IHH is rarely raised in the neonatal period, but it can be suspected in male infants presenting with cryptorchidism and/or microphallus [22, 23]. Cryptorchidism and microphallus are signs of undervirilization that may reflect an inadequate activation of the gonadotropic axis late in utero in IHH patients. However, in a series of 250 consecutive cases [7], these signs were detected only in a small minority of patients. This observation suggests that maternal hCG secretion alone is usually sufficient to stimulate endogenous testosterone production in male fetuses affected by the disease.

Therefore, most congenital forms of IHH are suspected at the time of normal puberty, in patients presenting with pubertal delay. The male-to-female ratio of affected patients is approximately 4:1. Diagnostic criteria for this syndrome include [24]: (1) Demonstration of prepubertal plasma concentrations of testosterone in boys and estradiol in girls. (2) Low or normal plasma LH and FSH concentrations (usually <4–5 IU/l), rather than the elevated concentrations expected with primary gonadal failure. (3) Otherwise normal anterior pituitary function (normal basal or stimulated plasma concentrations of thyroid and adrenal hormones as well as normal plasma prolactin and GH concentrations). (4) Normal anatomy of the hypothalamo-pituitary region on MRI scanning.

The most difficult (and almost impossible at a relatively young age) differential diagnosis of IHH is constitutional delay of puberty. Some authors have proposed the use of gonadotropin stimulation with exogenous GnRH to help in this differential diagnosis [25–28]. However, clinicians have more than often no choice but to rely upon time and clinical evolution to ascertain the diagnosis. Finally, the existence of adult-onset IHH, occurring after normal pubertal development and sexual maturation, deserves to be mentioned [29].

Three types of Mendelian inheritance have been described for the familial cases of Kallmann's syndrome: autosomal dominant, autosomal recessive and X-linked (associated with KAL-1 and DAX-1 mutations, see below), the latter form being the least frequent by far [21]. However, most reported cases are sporadic. And even so, only a very limited number of such sporadic cases bear a mutation in KAL-1 [30], confirming that the X-linked form is probably the least frequent. The identification of KAL-1 was made simultaneously by Legouis et al. [31] and by Franco et al. [32] in 1991. The physiology of this gene is reviewed extensively by Hardelin in another chapter of the present book. The quest for the autosomal gene(s) involved in the syndrome is underway in several laboratories throughout the world. Until now, it has turned out to be an extremely frustrating one for

Table 1. Summary of all mutations affecting genes of the hypothalamo-pituitary-gonadal axis in the human and resulting in hypogonadism and/or infertility: these mutations are found at all levels of that axis, and most of them have also been associated with a milder phenotypic expression than listed in the table (for further details, please refer to text)

Hypothalamus		Anterior pituitary		Ovary		Testes	
gene	phenotype	gene	phenotye	gene	phenotype	gene	phenotype
Kal-1	HH Anosmia	*GnRH-R*	Partial/ complete HH	*FSH-R*	1°/2° amenorrhea Delayed puberty	*FSH-R*	Defective spermatogenesis
Dax-1	HH AHC	*Dax-1*	HH AHC	*LH-R*	Normal puberty Amenorrhea	*LH-R*	Pseudo- hermaphroditism/ genital ambiguity
PC-1	HH 1° amenorrhea Obesity	*FSH-β*	1° amenorrhea Defective spermatogenesis			*DAX-1*	Defective spermatogenesis
Lep	HH Obesity	*LH-β*	Delayed puberty			*AR*	Androgen insensitivity
Lep-R	HH Obesity Short stature					*DAZ* *RBM*	OTA OTA

AR = Androgen receptor; DAZ = deleted in azoospermia; RBM = RNA binding motif protein; OTA = oligoteratoazoospermia.

the clinicians and researchers involved. The efforts of interested clinical researchers are now being funneled into collaborative networks and bear the promise of the identification of novel genes participating to the pathophysiology of Kallmann's syndrome.

Single-Gene Mutations and Hypogonadism

In recent years, a number of single-gene mutations resulting in human hypogonadism have been identified [for review, see 5, 6, 33]. These mutations can affect all levels of the hypothalamo-pituitary-gonadotropic axis and represent a model of single-gene diseases in humans (table 1). Many of these mutations have been described initially in rare patients presenting with a complete phenotype of IHH. However, it is becoming increasingly apparent that partial phenotypes of IHH can be linked to such single-gene mutations, and could account for more widespread cases. Mutations that have been associated with IHH in the human are reviewed below. In addition, a number of transcription factors which, when mutated, are associated with hypogonadism and various anterior pituitary defects

Table 2. Transcription factors involved in anterior pituitary development or function that have been associated with various forms of combined anterior pituitary dysfunction: for the most part, these transcription factors have been identified in rodents; genes associated with pituitary dysfunction in the human are indicated

Gene	Anterior pituitary	
	Human	Mouse
HESX-1	+	+
PROP-1/df	+	+
FOXC1	+	+
Pax6	–	+
Pitx1	–	+
POUF1/Pit1	+	+
LIF	–	+
Egr1	–	+

have been identified (table 2). At present, most of them have been found in rodent models, but their human equivalents are being gradually uncovered.

The Case of the Transcription Factor DAX-1

DAX-1 (for Dosage-sensitive sex reversal, Adrenal hypoplasia congenita critical region on the X chromosome gene 1 [34]) is a transcription factor of primordial importance in reproductive biology [35, 36]. It is co-expressed with SF-1 (for Steroidogenic Factor 1), another transcription factor (see below and Hanely et al. [37]), at all levels of the hypothalamo-pituitary-gonadal axis. Both nuclear hormone receptors follow a very similar pattern of developmental expression [38], and indeed DAX-1 and SF-1 seem to interact closely in several tissues to regulate the expression of target genes [39,40]. In most studies, DAX-1 behaves as a repressor of the transcriptional activity stimulated by SF-1 [41]. It is acting in a dosage-sensitive manner: duplication of this gene located on the short arm of the X chromosome results in persistence of Müllerian structures and XY sex reversal, whereas its absence is associated with male hypogonadism.

A relatively large number of mutations of DAX-1 have already been described in humans. Most of these mutations result in a C-terminally truncated protein, which therefore lacks a region that is critical for DAX-1 function [41].

The human disorder associated with DAX-1 inactivating mutations is transmitted as an X-linked trait affecting male patients hemizygous for the mutation. It associates the X-linked form of adrenal hypoplasia congenita (AHC) and hypogonadotropic hypogonadism [42, 43], the latter becoming apparent at the time of normal puberty. There is a wide range of phenotypic expression of DAX-1 mutations, from adrenal crisis in infancy and complete absence of pubertal development later in life [43, 44], to mild adult-onset adrenal insufficiency and defective sperm production [45].

This latter expression most closely resembles what has been observed in male mice bearing an inactivation of DAX-1, which exhibit defective spermatogenesis [46]. Of note, DAX-1–/– female mice are phenotypically normal [46]. In the human, heterozygous female carriers of DAX-1 mutations are also phenotypically normal, although the case of a hypogonadic female patient homozygous for a DAX-1 mutation has been reported [47]. With respect to the hypogonadotropic hypogonadism present in the vast majority of patients studied to date, there is evidence for a hypothalamic [48, 49] as well as a pituitary level for the defect [50–52], but no good genotype-phenotype correlation could be drawn until now.

The Other Transcription Factor, SF-1

An insight into the biological role played by SF-1 in mammals was initially provided by the generation of a mouse bearing an inactivation of SF-1 [53], which displayed several endocrine defects reminiscent of the phenotype of human patients with DAX-1 mutations: adrenal and gonadal agenesis, as well as defective biosynthesis and secretion of gonadotropins. In addition, SF-1–/– mice exhibit complete XY sex reversal. This phenotype is consistent with the role of SF-1 in regulating the biosynthesis of androgens and the expression of MIS (Müllerian-inhibiting substance) as well as that of gonadotropins [37]. More recently, 2 cases of human SF-1 mutations were reported [54, 55]. The first subject was a phenotypically female, 46,XY patient who presented with primary adrenal failure at 2 weeks of age. An initial presumptive diagnosis of congenital lipoid adrenal hyperplasia was made. Further evaluation at age 10 years disclosed a normal LH response to GnRH stimulation, elevated baseline FSH levels and no testosterone response to hCG [55]. At laparotomy, normal Müllerian structures but streak-like gonads were found, and removed. This finding led to the questioning of the diagnosis of StAR deficiency: upon direct sequencing of the SF-1 gene, a heterozygous mutation was found in a portion of the protein corresponding to its DNA-binding domain [55]. Later, transdermal estradiol treatment effectively induced breast development and uterus enlargement, followed by normal menstruations upon introduction of cyclical progesterone.

The second case published was that of a 46,XX, phenotypically normal girl, presenting with signs and symptoms of adrenal insufficiency at age 14 months

[54]. The biochemical picture was consistent with primary adrenal insufficiency, and congenital adrenal hyperplasia was excluded by the proper biochemical work-up. Subsequently, she did well under mineralo- and glucocorticoid hormone replacement therapy. Further laboratory evaluation was performed at age 27 months. It included baseline LH and FSH measurements which were in the normal range for age, assessment of inhibin A and B levels which were also normal for her age, and an abdominal MRI which disclosed absent adrenal glands but apparently normal ovaries at the correct location [54]. The results of this work-up suggest that ovarian development and function are normal in this particular patient. Mutational analysis of SF-1 revealed the existence of a de novo heterozygous missense mutation in a conserved region of the protein, which interferes with its DNA binding in vitro [54].

Taken together, these two reports confirm the dependence of human adrenal gland development and function upon SF-1. In addition, the finding of complete XY sex reversal in an affected male demonstrates the crucial role of SF-1 in normal testicular differentiation. In contrast, the normal phenotype of the gonadotrope axis, including apparently normal ovaries, in a prepubertal girl with an SF-1 mutation probably indicates that SF-1 is not an ovary-determining gene. It remains however to be established whether SF-1 will be necessary for normal pubertal development in that girl. Indeed, the possibility that SF-1 will be required for proper GnRH and/or gonadotropin biosynthesis and release at the time of puberty remains open, even though gonadotropin secretion was preserved in the affected male patient.

Gonadotropin-Releasing Hormone Receptor

In contrast to GnRH (see below), the almost universally adequate response of IHH patients to pulsatile GnRH administration seemed for a long time to preclude the inclusion of the GnRH receptor (GnRH-R) in the list of candidate genes for this syndrome [56]. However, seven different kindreds bearing inactivating mutations of the GnRH-R have now been described [57–63] (fig. 1), identifying this receptor as the first autosomal gene associated with IHH. Their phenotypic expression ranges from complete hypogonadism with cryptorchid testes [59] to mild pubertal retardation and near normal development of secondary sexual characteristics [58]. Two mutations associated in the compound heterozygous state (Q106R in the first extracellular loop and R262Q in the third intracellular loop) have been found in the majority of affected patients. In vitro studies have confirmed that all the mutations described to date impair signal transduction to some degree.

There is a general correlation between the magnitude of transduction impairment and the clinical phenotype: patients with completely inactivating mutations do not respond to pharmacological stimulation with exogenous GnRH. This is

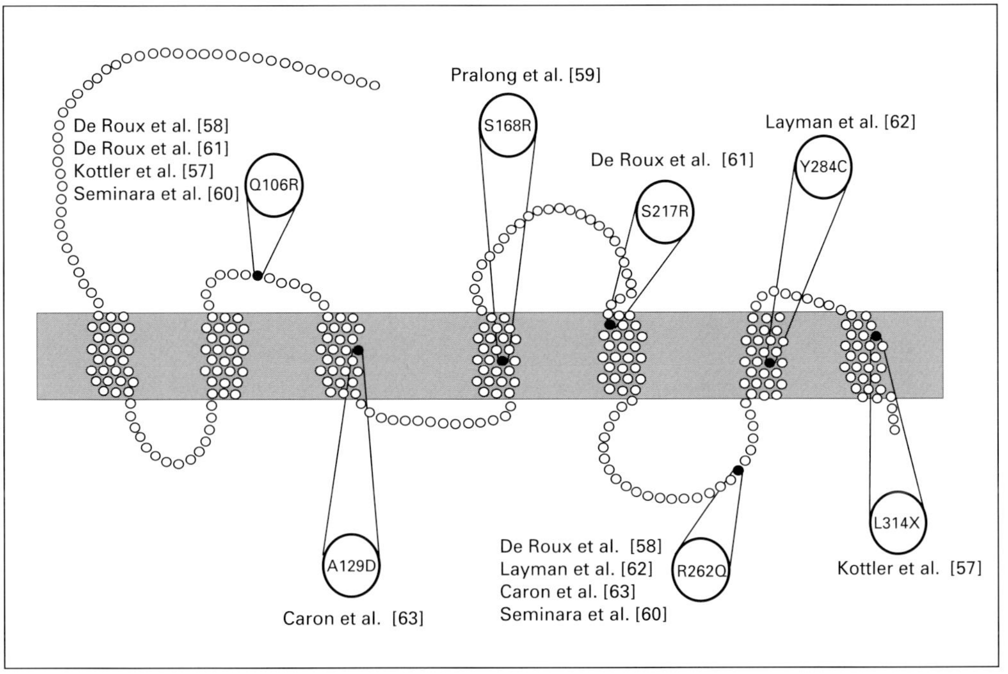

Fig. 1. Schematic representation of the GnRH-R. All loss-of-function mutations of the GnRH-R gene reported to date are indicated. They are inherited as an autosomal recessive trait. Most of these mutations were present in the compound heterozygous state (for further details, see text).

exemplified by the only patient bearing a homozygous mutation (S168R), who displayed complete IHH with an absolute resistance to GnRH stimulation [59]. The S168R mutation identified in this patient was found to completely abolish ligand binding in vitro (fig. 2), demonstrating the importance for proper ligand binding of this serine residue located within the fourth transmembrane domain of the receptor [59]. However, the partially inactivating mutations display a more variable phenotypic expression [61, 63]. Indeed, identical mutations present in the compound heterozyogte state in different family members have been associated with different levels of spontaneous activation of the gonadotrope axis [61, 63]. These studies have nevertheless allowed to confirm the importance of the third transmembrane domain of the GnRH-R for receptor activation [61, 63]. This region has been associated with ligand binding in other G-protein-coupled receptors [64], members of the seven transmembrane family of receptors [65].

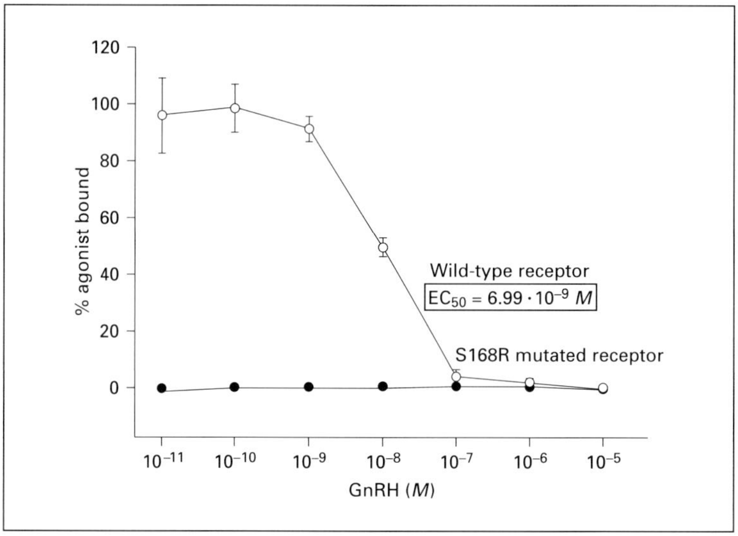

Fig. 2. COS-7 cells were transfected with vectors expressing the wild-type or S168R mutated GnRH-R. Intact cell receptor binding was measured by incubating cells with different concentrations of (125)I-(*D*Trp6(N-Et)Pro9,DesGly10)-GnRH. Cells transfected with the cDNA encoding the wild-type receptor displayed specific binding of the GnRH analogue that could be displaced with unlabelled competitor. Cells transfected with the cDNA encoding the muttant receptor displayed complete lack of binding [59].

Because of this wide range of phenotypic expression, it can be hypothesized that patients with 'stalled' puberty, or adult patients with partial pubertal development, could harbor GnRH-R mutations more frequently than previously expected. Further studies of the GnRH-R gene should be directed at genotyping such patients presenting as 'intermediate' phenotypes.

Loss-of-Function Mutations of the Gonadotropin Receptor Genes

Unlike the other genes discussed here, loss-of-function mutations in the gonadotropin receptor genes result in hypergonadotropic hypogonadism [for review, see 66, 67] whereas mutations in their ligands, LH and FSH, result in the rare cases of isolated gonadotropin deficiency [68–71]. A comprehensive description of the phenotypes associated with such mutations is beyond the scope of the present chapter. However, some features presented by these patients deserve to be stressed.

One is the gender dependency of the phenotypic expression of inactivating mutations of the LH receptor. Female patients usually have hypergonadotropic hypogonadism associated with normal secondary sexual characteristics [66], whereas male patients present with a variable degree of genital ambiguity depending on the degree of functional impairment of the receptor [66, 72]. The exceptional case of a patient with an LH receptor mutation presenting as delayed puberty and azoospermia was recently reported [73]. The normal sexual differentiation and the appropriate response of testosterone to hCG stimulation in this patient in whom a large part of exon 10 of the LH receptor was deleted argues for a differential activation of the receptor by LH and chorionic gonadotropin [73]. Taken together, all these reports demonstrate the physiological importance of the LH-LH receptor system for normal male sex differentiation and spermatogenesis, and for fertility in the female.

A gender-specific expression of loss-of-function mutations has also been found in the case of the FSH receptor. In female patients, they are associated with primary amenorrhea and a variable degree of delayed pubertal development. The first missense mutation of the FSH receptor, located in the extracellular domain of the protein and leading to impaired cell surface expression, was initially reported in hypergonadotropic hypogonadic Finnish women [74]. Ovarian dysgenesis was found in many of these women, but some degree of follicular maturation was nevertheless observed in others [74]. Five male family members homozygous for the same mutation were subsequently identified [75]: they were all normally virilized and had normal testosterone levels with high FSH and normal or high LH levels. Surprisingly, azoospermia was not the rule, some of these men exhibiting only a moderate degree of teratozoospermia. Two of these patients had even fathered 2 children. These findings have been challenging for the concept that FSH is the critical hormone of spermatogenesis, especially since they corroborate almost exactly the phenotype observed in mice deficient for either FSH-β [76] or the FSH receptor [77]. In contrast, in a male patient with an FSH-β mutation, puberty had occurred but there was no spermatozoa in the ejaculate and testicular biopsy revealed spermatogenic arrest [70]. In conclusion, these human data suggest that FSH is crucial for normal follicular development in females. In males, FSH may not constitute an absolute requirement for the initiation and maintenance of spermatogenesis, but it is needed for optimal sperm production. However, the hypothesis that mutations in the FSH receptor could be widely responsible for the syndrome of 'idiopathic' male infertility could not be confirmed [78].

Despite the peculiar phenotypes observed in these cases [70, 75], hypogonadotropic hypogonadism in its more classical form remains one of the very few treatable causes of defective spermatogenesis and male infertility. Pulsatile GnRH therapy and exogenous gonadotropin treatment have indeed proven extremely successful to achieve pubertal development, and to initiate as well as sustain effective sperm production in affected patients.

Table 3. Leptin and leptin-receptor deficiency are associated with hypogonadotropic hypogonadism in the human

	OB 1 Montague et al. [90]	OB 2 Montague et al. [90]	OB 3 Strobel et al. [91]	OB 4 Strobel et al. [91]	OB-R 1 Clément et al. [92]	OB-R 2 Clément et al. [92]
Age at diagnosis, years	8	2	34	22	19	19
Sex	F	M	F	M	F	F
Mutation	G deletion at codon 133 (frameshift)	G deletion at codon 133 (frameshift)	R105W	R105W	G to A in splice donor site, exon 16	G to A in splice donor site, exon 16
Clinical features	prepubertal	prepubertal	primary amenorrhea	delayed puberty	primary	primary
LH, IU/l	<0.2	<0.2	NA	4.4	<0.2	<0.8
FSH, IU/l	0.8	0.2	NA	9.0	<0.1	1.2
E_2, pmol/l	<20	–	NA	–	17	13
T, nmol/l	–	<0.2	–	5	–	–

OB = Patients with leptin mutations; OB-R = patients with leptin receptor mutations. All mutations were present in the homozygote state in the affected patients.

Leptin and the Leptin Receptor

The 167-amino-acid protein leptin is a secreted product of the adipocyte initially isolated in the *Lep^{ob}* mouse. This mouse displays an autosomal recessive syndrome of obesity secondary to marked hyperphagia, associated with hypogonadotropic hypogonadism as well as other neuroendocrine and metabolic abnormalities [79, 80]. Following the isolation of the *Ob* gene encoding leptin [81], it was demonstrated that the phenotype of the *Lep^{ob}* mouse was due to an inactivating mutation of leptin. Indeed, administration of the recombinant peptide to *Lep^{ob}* mice could correct both their hyperphagia [82–84] and their hypogonadism [85, 86]. This early work [82–84] demonstrated that one important function of leptin is to signal the nutritional status of the organism to the hypothalamus, where it acts as a satiety factor. In addition, it was later suggested that leptin could be an important modulator of several neuroendocrine functions including reproductive activity [87]. All these effects depend on the binding of leptin to its cognate receptor, OB-R [88], which is expressed by several hypothalamic neuronal populations.

Soon after the discovery of leptin, it became apparent that human obesity does not result from leptin deficiency [89]. However, exceptional cases of human leptin [90, 91] or leptin receptor [92] mutations are associated with both obesity and hypogonadotropic hypogonadism. In the first published family presenting a leptin mutation [90], the 2 affected patients (2 first-degree cousins) were prepubertal at the time of diagnosis, hence rendering the diagnosis of hypogonadism impossible (see patients OB 1 and OB 2, table 3). However, another family with 2 affected adult individuals (patients OB 3 and OB 4, table 3) was described more recently [91]. These 2 patients, a male and a female, presented morbid hyperphagic obesity as well as complete impuberism secondary to hypogonadotropic hypogonadism. This observation is entirely consistent with the phenotype of the *Lep^ob^* mouse, and suggests that leptin plays a role in human puberty and reproduction. This hypothesis was confirmed by the study of members of a French kindred harboring a loss-of-function mutation in the leptin receptor gene [92]. Two of the 3 affected patients were old enough at diagnosis to fulfill the clinical criteria of IHH (patients OB-R 1 and OB-R 2, table 3), further demonstrating the physiological importance of leptin for normal human puberty.

Very recently, the results of leptin replacement therapy in human leptin deficiency became available [93]. In this study, recombinant leptin administration to a 12-year-old leptin-deficient girl (patient OB 1, table 3) induced a dramatic and sustained (over almost a year of treatment) weight loss. Furthermore, it was found to restore the nighttime physiologic pulsatile LH secretion usually seen in a normal girl of that age. Taken together, all these observations demonstrate that as is the case in rodents, leptin definitively participates to normal human puberty via effects at the hypothalamic level to stimulate GnRH secretion, although its precise mechanism of action in the initiation of puberty remains to be defined.

The Particular Case of GnRH

Because of the crucial role played by the decapeptide GnRH in gonadotrope axis activation, it was deemed very early on that it would be a very logical candidate gene of IHH. This hypothesis was initially corroborated by the genetic study of the hypogonadal (*hpg*) mouse [94]. The *hpg* mouse is a rodent model of hypogonadotropic hypogonadism resulting from a spontaneous inactivating mutation of the GnRH gene [94]. Therefore, molecular studies looking for mutations in the human GnRH gene were performed in the early 1990s, in small series of patients presenting a 'classical' (complete) phenotype of IHH [95, 96]. However, all patients studied turned up to have a normal GnRH gene sequence. These results, coupled with the almost universally positive response of IHH patients to pulsatile GnRH therapy, yielded the notion that isolated GnRH deficiency rather represents 'functional' deficiency. The elucidation of the role of KAL-1 in Kallmann's syndrome further confirmed this notion, by providing a mechanism for the functional insufficiency: affected patients can synthesize and probably secrete a normal GnRH protein, but their GnRH neurons are not properly located in the basal hypothalamus, and therefore GnRH does not reach the gonadotrope cells of the anterior pituitary [97, 98].

Despite the negative results of these two early screenings [95, 96], GnRH remains a valid candidate gene of IHH. Based on the description of the *hpg* mouse as well as the phenotype of Kallmann's syndrome patients, a subject with a GnRH gene mutation would be expected to have a 'classical' phenotype of IHH with normal sexual differentiation, probably no anosmia, and complete failure of pubertal development.

Conclusions

Although it is only beginning, the elucidation of the molecular basis of hypogonadism has already allowed to uncover patients with very different and sometimes unanticipated phenotypes. The study of these patients is providing deeper insight into the central regulation of sexual differentiation and maturation, and is shedding new light upon the pathophysiology of hypogonadism. It is now clear that the two clinical entities constituted by Kallmann's syndrome and IHH cover a broader range of phenotypic expression than previously thought. This finding has led to the hypothesis that milder phenotypes may exist, and be more common than previously acknowledged. Until now, this flow of novel information has not altered the day-to-day management of these patients, but it might well do so in the future, when more advanced diagnostic and therapeutic techniques including gene therapy become available.

Acknowledgments

The author is grateful to Dr. Fulgencio Gomez for stimulating discussions. F.P. Pralong is the recipient of a Research Development Career Award from the Prof. Dr. Max Cloëtta Foundation. This work was supported in part by a grant from the Swiss National Science Foundation (grant 2-059 465.99).

References

1 Marshall JC, Kelch RP: Gonadotropin-releasing hormone: Role of pulsatile secretion in the regulation of reproduction. N Engl J Med 1986;315:1459–1468.
2 Santen RJ, Bardin CW: Episodic luteinizing hormone secretion in man. Pulse analysis, clinical interpretation, physiologic mechanisms. J Clin Invest 1973;52:2617–2628.
3 Clarke IJ, Cummins JT: The temporal relationship between gonadotropin-releasing hormone and luteinizing hormone secretion in ovariectomized ewes. Endocrinology 1982;111:1737–1739.
4 Crowley WF Jr, Filicori M, Spratt DI, Santoro NF: The physiology of gonadotropin-releasing hormone secretion in men and women. Recent Prog Horm Res 1985;41:473–531.
5 Achermann JC, Jameson JL: Fertility and infertility: Genetic contributions from the hypothalamic-pituitary-gonadal axis. Mol Endocrinol 1999;13:812–818.

6 Adashi EY, Hennebold JD: Single-gene mutations resulting in reproductive dysfunction in women. N Engl J Med 1999;340:709–718.
7 Seminara SB, Hayes FJ, Crowley WF Jr: Gonadotropin-releasing hormone deficiency in the human (idiopathic hypogonadotropic hypogonadism and Kallmann's syndrome): Pathophysiological and genetic considerations. Endocr Rev 1998;19:521–539.
8 Maestre de San Juan: Teratologia: falta total de los nervios olfactorios con anosmia en un individio en quien existia un atrofia congenita de los testiculos y miembro viril. El Siglo Med 1856;3:211–221.
9 Kallmann FJ, Schoenfeld WA: The genetic aspects of primary eunuchoidism. Am J Ment Defic 1944;158:203–236.
10 De Morsier G: Etudes sur les dysraphies crânio-encéphaliques. 1. Agénésie des lobes olfactifs (telencephaloschizis lateral) et des commissures calleuses et antérieures (telencephaloschizis median). La dysplasie olfacto-génitale. Schweiz Arch Nerurol Neurochir Psychiatr 1954;74:309.
11 Vogl TJ, Stemmler J, Heye B, Schopohl J, Danek A, Bergman C, Balzer JO, Felix R: Kallman syndrome versus idiopathic hypogonadotropic hypogonadism at MR imaging. Radiology 1994;191:53–57.
12 Yousem DM, Turner WJ, Li C, Snyder PJ, Doty RL: Kallmann syndrome: MR evaluation of olfactory system. AJNR Am J Neuroradiol 1993;14:839–843.
13 Klingmuller D, Dewes W, Krahe T, Brecht G, Schweikert HU: Magnetic resonance imaging of the brain in patients with anosmia and hypothalamic hypogonadism (Kallmann's syndrome). J Clin Endocrinol Metab 1987;65:581–584.
14 Quinton R, Duke VM, de Zoysa PA, Platts AD, Valentine A, Kendall B, Pickman S, Kirk JM, Besser GM, Jacobs HS, Bouloux PM: The neuroradiology of Kallmann's syndrome: A genotypic and phenotypic analysis. J Clin Endocrinol Metab 1996;81:3010–3017.
15 Filippi G: Klinefelter's syndrome in Sardinia. Clinical report of 265 hypogonadic males detected at the time of military check-up. Clin Genet 1986;30:276–284.
16 Fromantin M, Gineste J, Didier A, Rouvier J: Les impubérismes et les hypogonadismes à l'incorporation. Etude statistique. Probl Actuels Endocrinol Nutr 1973;16:179–199.
17 Hermanussen M, Sippel W: Heterogeneity of Kallmann's syndrome. Clin Genet 1985;28:106–111.
18 Boyar RM, Wu RH, Kapen S, Hellman L, Weitzman ED, Finkelstein JW: Clinical and laboratory heterogeneity in idiopathic hypogonadotropic hypogonadism. J Clin Endocrinol Metab 1976;43:1268–1275.
19 Spratt DI, Carr DB, Merriam GR, Scully RE, Rao PN, Crowley WF Jr: The spectrum of abnormal patterns of gonadotropin-releasing hormone secretion in men with idiopathic hypogonadotropic hypogonadism: Clinical and laboratory correlations. J Clin Endocrinol Metab 1987;64:283–291.
20 Barkan AL, Reame NE, Kelch RP, Marshall JC: Idiopathic hypogonadotropic hypogonadism in men: Dependence of the hormone responses to gonadotropin releasing hormone (GnRH) on the magnitude of the endogenous GnRH secretory defect. J Clin Endocrinol Metab 1985;61:1118–1125.
21 Waldstreicher J, Seminara SB, Jameson JL, Geyer A, Nachtigall LB, Boepple PA, Holmes LB, Crowley WF Jr: The genetic and clinical heterogeneity of gonadotropin-releasing hormone deficiency in the human. J Clin Endocrinol Metab 1996;81:4388–4395.
22 Van Dop C, Burstein S, Conte FA, Grumbach MM: Isolated gonadotropin deficiency in boys: Clinical characteristics and growth. J Pediatr 1987;111:684–692.
23 Bardin CW, Ross GT, Rifkind AB, Cargille CM, Lipsett MB: Studies of the pituitary-Leydig cell axis in young men with hypogonadotropic hypogonadism and hyposmia: Comparison with normal men, prepuberal boys, and hypopituitary patients. J Clin Invest 1969;48:2046–2056.
24 Pralong FP, Crowley WF Jr: Delayed puberty: Diagnosis and treatment; in Rose B, Savin C (eds): UpToDate in Endocrinology and Diabetes, 2000, vol 8-2.
25 Lanes R, Gunczler P, Osuna JA, Palacios A, Carrillo E, Ramirez X, Garcia C, Paoli M, Villaroel O: Effectiveness and limitations of the use of the gonadotropin-releasing hormone agonist leuprolide acetate in the diagnosis of delayed puberty in males. Horm Res 1997;48:1–4.

26 Ghai K, Cara JF, Rosenfield RL: Gonadotropin-releasing hormone agonist (nafarelin) test to differentiate gonadotropin deficiency from constitutionally delayed puberty in teenage boys – A clinical research center study. J Clin Endocrinol Metab 1995;80:2980–2986.
27 Smals AG, Hermus AR, Boers GH, Pieters GF, Benraad TJ, Kloppenborg PW: Predictive value of luteinizing hormone releasing hormone (LHRH) bolus testing before and after 36-hour pulsatile LHRH administration in the differential diagnosis of constitutional delay of puberty and male hypogonadotropic hypogonadism. J Clin Endocrinol Metab 1994;78:602–608.
28 Partsch CJ, Hermanussen M, Sippell WG: Differentiation of male hypogonadotropic hypogonadism and constitutional delay of puberty by pulsatile administration of gonadotropin-releasing hormone. J Clin Endocrinol Metab 1985;60:1196–1203.
29 Nachtigall LB, Boepple PA, Pralong FP, Crowley WF Jr: Adult-onset idiopathic hypogonadotropic hypogonadism – A treatable form of male infertility. N Engl J Med 1997;336:410–415.
30 Georgopoulos NA, Pralong FP, Seidman CE, Seidman JG, Crowley WF Jr, Vallejo M: Genetic heterogeneity evidenced by low incidence of KAL-1 gene mutations in sporadic cases of gonadotropin-releasing hormone deficiency. J Clin Endocrinol Metab 1997;82:213–217.
31 Legouis R, Hardelin JP, Levilliers J, Claverie JM, Compain S, Wunderle V, Millasseau P, Le Paslier D, Cohen D, Caterina D, Bougueleret L, Delemarre-Van de Waal H, Lutfalla G, Weissenbach J, Petit C: The candidate gene for the X-linked Kallmann syndrome encodes a protein related to adhesion molecules. Cell 1991;67:423–435.
32 Franco B, Guioli S, Pragliola A, Incerti B, Bardoni B, Tonlorenzi R, Carrozzo R, Maestrini E, Pieretti M, Taillon-Miller P, Brown CJ, Willard HF, Lawrence C, Persico MG, Camerino G, Ballabio A: A gene deleted in Kallmann's syndrome shares homology with neural cell adhesion and axonal path-finding molecules. Nature 1991;353:529–536.
33 Seminara SB, Oliveira LM, Beranova M, Hayes FJ, Crowley WF Jr: Genetics of hypogonadotropic hypogonadism. J Endocrinol Invest 2000;23:560–565.
34 Walker AP, McCabe ERB, Meitinger T, Monaco AP, Sassone-Corsi P, Camerino G: An unusual member of the nuclear hormone receptor superfamily responsible for X-linked adrenal hypoplasia congenita. Nature 1994;372:635–641.
35 Muscatelli F, Strom TM, Walker AP, Zanaria E, Recan D, Meindl A, Bardoni B, Guioli S, Zehetner G, Rabl W, Schwartz HP, Kaplan JC, Camerino G, Meitinger T, Monaco AP: Mutations in the DAX-1 gene give rise to both X-linked adrenal hypoplasia congenita and hypogonadotropic hypogonadism. Nature 1994;372:672–676.
36 Guo W, Burris TP, McCabe ER: Expression of DAX-1, the gene responsible for X-linked adrenal hypoplasia congenita and hypogonadotropic hypogonadism, in the hypothalamic-pituitary-adrenal/gonadal axis. Biochem Mol Med 1995;56:8–13.
37 Hanley NA, Ikeda Y, Luo X, Parker KL: Steroidogenic factor 1 (SF-1) is essential for ovarian development and function. Mol Cell Endocrinol 2000;163:27–32.
38 Ikeda Y, Swain A, Weber TJ, Hentges KE, Zanaria E, Lalli E, Tamai KT, Sassone-Corsi P, Lovell-Badge R, Camerino G, Parker KL: Steroidogenic factor 1 and Dax-1 colocalize in multiple cell lineages: Potential links in endocrine development. Mol Endocrinol 1996;10:1261–1272.
39 Burris TP, Guo W, Le T, McCabe ER: Identification of a putative steroidogenic factor-1 response element in the DAX-1 promoter. Biochem Biophys Res Commun 1995;214:576–581.
40 Yu RN, Ito M, Jameson JL: The murine Dax-1 promoter is stimulated by SF-1 (steroidogenic factor-1) and inhibited by COUP-TF (chicken ovalbumin upstream promoter-transcription factor) via a composite nuclear receptor-regulatory element. Mol Endocrinol 1998;12:1010–1022.
41 Ito M, Yu R, Jameson JL: DAX-1 inhibits SF-1-mediated transactivation via a carboxy-terminal domain that is deleted in adrenal hypoplasia congenita. Mol Cell Biol 1997;17:1476–1483.
42 Golden MP, Lippe BM, Kaplan SA: Congenital adrenal hypoplasia and hypogonadotropic hypogonadism. Am J Dis Child 1977;131:1117–1118.
43 Prader A, Zachmann M, Illig R: Luteinizing hormone deficiency in hereditary congenital adrenal hypoplasia. J Pediatr 1975;86:421–422.
44 Uttley WS: Familial congenital adrenal hypoplasia. Arch Dis Child 1968;43:724–730.
45 Tabarin A, Achermann JC, Recan D, Bex V, Bertagna X, Christin-Maitre S, Ito M, Jameson JL, Bouchard P: A novel mutation in DAX1 causes delayed-onset adrenal insufficiency and incomplete hypogonadotropic hypogonadism. J Clin Invest 2000;105:321–328.

46 Yu RN, Ito M, Saunders TL, Camper SA, Jameson JL: Role of Ahch in gonadal development and gametogenesis. Nat Genet 1998;20:353–357.

47 Merke DP, Tajima T, Baron J, Cutler GB Jr: Hypogonadotropic hypogonadism in a female caused by an X-linked recessive mutation in the DAX1 gene. N Engl J Med 1999;340:1248–1252.

48 Kruse K, Sippell WG, Schnakenburg KV: Hypogonadism in congenital adrenal hypoplasia: Evidence for a hypothalamic origin. J Clin Endocrinol Metab 1984;58:12–17.

49 Partsch CJ, Sippell WG: Hypothalamic hypogonadism in congenital adrenal hypoplasia. Horm Metab Res 1989;21:623–625.

50 Gordon D, Cohen HN, Beastall GH, Hay ID, Thomson JA: Contrasting effects of subcutaneous pulsatile GnRH therapy in congenital adrenal hypoplasia and Kallmann's syndrome. Clin Endocrinol 1984;21:597–603.

51 Bovet P, Reymond MJ, Rey F, Gomez F: Lack of gonadotropic response to pulsatile gonadotropin-releasing hormone in isolated hypogonadotropic hypogonadism associated to congenital adrenal hypoplasia. J Endocrinol Invest 1988;11:201–204.

52 Habiby RL, Boepple P, Nachtigall L, Sluss PM, Crowley WF Jr, Jameson JL: Adrenal hypoplasia congenita with hypogonadotropic hypogonadism: Evidence that DAX-1 mutations lead to combined hypothalamic and pituitary defects in gonadotropin production. J Clin Invest 1996;98:1055–1062.

53 Luo X, Ikeda Y, Parker KL: A cell-specific nuclear receptor is essential for adrenal and gonadal development and sexual differentiation. Cell 1994;77:481–490.

54 Biason-Lauber A, Schoenle EJ: Apparently normal ovarian differentiation in a prepubertal girl with transcriptionally inactive steroidogenic factor 1 (NR5A1/SF-1) and adrenocortical insufficiency. Am J Hum Genet 2000;67:1563–1568.

55 Achermann JC, Ito M, Hindmarsh PC, Jameson JL: A mutation in the gene encoding steroidogenic factor-1 causes XY sex reversal and adrenal failure in humans. Nat Genet 1999;22:125–126.

56 Whitcomb RW, Crowley WF Jr: Clinical review 4: Diagnosis and treatment of isolated gonadotropin-releasing hormone deficiency in men. J Clin Endocrinol Metab 1990;70:3–7.

57 Kottler ML, Chauvin S, Lahlou N, Harris CE, Johnston CJ, Lagarde JP, Bouchard P, Farid NR, Counis R: A new compound heterozygous mutation of the gonadotropin-releasing hormone receptor (L314X, Q106R) in a woman with complete hypogonadotropic hypogonadism: Chronic estrogen administration amplifies the gonadotropin defect. J Clin Endocrinol Metab 2000;85:3002–3008.

58 De Roux N, Young J, Misrahi M, Genet R, Chanson P, Schaison G, Milgrom E: A family with hypogonadotropic hypogonadism and mutations in the gonadotropin-releasing hormone receptor. N Engl J Med 1997;337:1597–1602.

59 Pralong FP, Gomez F, Castillo E, Cotecchia S, Abuin L, Aubert ML, Portmann L, Gaillard RC: Hypogonadotropic hypogonadism associated with a novel inactivating mutation of the gonadotropin-releasing hormone receptor. J Clin Endocrinol Metab 1999;3811–3816.

60 Seminara SB, Beranova M, Oliveira LM, Martin KA, Crowley WF Jr, Hall JE: Successful use of pulsatile gonadotropin-releasing hormone (GnRH) for ovulation induction and pregnancy in a patient with GnRH receptor mutations. J Clin Endocrinol Metab 2000;85:556–562.

61 De Roux N, Young J, Brailly-Tabard S, Misrahi M, Milgrom E, Schaison G: The same molecular defects of the gonadotropin-releasing hormone receptor determine a variable degree of hypogonadism in affected kindred. J Clin Endocrinol Metab 1999;84:567–572.

62 Layman LC, Cohen DP, Jin M, Xie J, Li Z, Reindollar RH, Bolbolan S, Bick DP, Sherins RR, Duck LW, Musgrove LC, Sellers JC, Neill JD: Mutations in gonadotropin-releasing hormone receptor gene cause hypogonadotropic hypogonadism. Nat Genet 1998;18:14–15.

63 Caron P, Chauvin S, Cristin-Maitre S, Bennet A, Lahlou N, Counis R, Bouchard P, Kottler ML: Resistance of hypogonadic patients with mutated GnRH receptor genes to pulsatile GnRH administration. J Clin Endocrinol Metab 1999;84:990–996.

64 Collu R, Tang J, Castagne J, Lagace G, Masson N, Huot C, Deal C, Delvin E, Faccenda E, Eidne KA, Van Vliet G: A novel mechanism for isolated central hypothyroidism: Inactivating mutations in the thyrotropin-releasing hormone receptor gene. J Clin Endocrinol Metab 1997;82:1561–1565.

65 Probst WC, Snyder LA, Schuster DI, Brosius J, Sealfon SC: Sequence alignment of the G-protein coupled receptor superfamily. DNA Cell Biol 1992;11:1–20.

66 Beck-Peccoz P, Romoli R, Persani L: Mutations of LH and FSH receptors. J Endocrinol Invest 2000;23:566–572.

67 Huhtaniemi I: The Parkes lecture. Mutations of gonadotrophin and gonadotrophin receptor genes: What do they teach us about reproductive physiology? J Reprod Fertil 2000;119:173–186.

68 Weiss J, Axelrod L, Whitcomb RW, Harris PE, Crowley WF, Jameson JL: Hypogonadism caused by a single amino acid substitution in the beta subunit of luteinizing hormone. N Engl J Med 1992; 326:179–183.

69 Phillip M, Arbelle JE, Segev Y, Parvari R: Male hypogonadism due to a mutation in the gene for the beta-subunit of follicle-stimulating hormone. N Engl J Med 1998;338:1729–1732.

70 Lindstedt G, Nystrom E, Matthews C, Ernest I, Janson PO, Chatterjee K: Follitropin (FSH) deficiency in an infertile male due to FSH-β gene mutation. A syndrome of normal puberty and virilization but underdeveloped testicles with azoospermia, low FSH but high lutropin and normal serum testosterone concentrations. Clin Chem Lab Med 1998;36:663–665.

71 Layman LC, Lee EJ, Peak DB, Namnoum AB, Vu KV, van Lingen BL, Gray MR, McDonough PG, Reindollar RH, Jameson JL: Delayed puberty and hypogonadism caused by mutations in the follicle-stimulating hormone beta-subunit gene. N Engl J Med 1997;337:607–611.

72 Kremer H, Kraaij R, Toledo SP, Post M, Fridman JB, Hayashida CY, van Reen M, Milgrom E, Ropers HH, Mariman E, et al: Male pseudohermaphroditism due to a homozygous missense mutation of the luteinizing hormone receptor gene. Nat Genet 1995;9:160–164.

73 Gromoll J, Eiholzer U, Nieschlag E, Simoni M: Male hypogonadism caused by homozygous deletion of exon 10 of the luteinizing hormone (LH) receptor: Differential action of human chorionic gonadotropin and LH. J Clin Endocrinol Metab 2000;85:2281–2286.

74 Aittomaki K, Lucena JL, Pakarinen P, Sistonen P, Tapanainen J, Gromoll J, Kaskikari R, Sankila EM, Lehvaslaiho H, Engel AR, Nieschlag E, Huhtaniemi I, de la Chapelle A: Mutation in the follicle-stimulating hormone receptor gene causes hereditary hypergonadotropic ovarian failure. Cell 1995;82:959–968.

75 Tapanainen JS, Aittomaki K, Min J, Vaskivuo T, Huhtaniemi IT: Men homozygous for an inactivating mutation of the follicle-stimulating hormone receptor gene present variable suppression of spermatogenesis and fertility. Nat Genet 1997;15:205–206.

76 Kumar TR, Wang Y, Lu N, Matzuk MM: Follicle-stimulating hormone is required for ovarian follicle maturation but not male fertility. Nat Genet 1997;15:201–204.

77 Abel MH, Wootton AN, Wilkins V, Huhtaniemi I, Knight PG, Charlton HM: The effect of a null mutation in the follicle-stimulating hormone receptor gene on mouse reproduction. Endocrinology 2000;141:1795–1803.

78 Simoni M, Gromoll J, Hoppner W, Kamischke A, Krafft T, Stahle D, Nieschlag E: Mutational analysis of the follicle-stimulating hormone (FSH) receptor in normal and infertile men: Identification and characterization of two discrete FSH receptor isoforms. J Clin Endocrinol Metab 1999;84: 751–755.

79 Garthwaite TL, Martinson DR, Tseng LF, Hagen TC, Menahan LA: A longitudinal hormonal profile of the genetically obese mouse. Endocrinology 1980;107:671–676.

80 Edwardson JA , Hough CAM: The pituitary-adrenal system of the genetically obese (*ob/ob*) mouse. J Endocrinol 1975;65:99–107.

81 Zhang Y, Proenca R, Maffei M, Barone M, Leopold L, Friedman JM: Positional cloning of the mouse obese gene and its human homologue. Nature 1994;372:425–432.

82 Pelleymounter MA, Cullen MJ, Baker MB, Hecht R, Winters D, Boone T, Collins F: Effects of the *obese* gene product on body weight regulation in *ob/ob* mice. Science 1995;269:540–543.

83 Halaas JL, Gajiwala KS, Maffei M, Cohen SL, Chait BT, Rabinowitz D, Lallone RL, Burley SK, Friedman JM: Weight-reducing effects of the plasma protein encoded by the *obese* gene. Science 1995;269:543–546.

84 Campfield LA, Smith FJ, Guisez Y, Devos R, Burn P: Recombinant mouse OB protein: Evidence for a peripheral signal linking adiposity and central neural networks. Science 1995;269:546–549.

85 Mounzih K, Lu R, Chehab FF: Leptin treatment rescues the sterility of genetically obese *ob/ob* males. Endocrinology 1997;138:1190–1193.

86 Chehab FF, Lim ME, Lu R: Correction of the sterility defect in homozygous obese female mice by treatment with the human recombinant leptin. Nat Genet 1996;12:318–320.

87 Ahima RS, Prabakaran D, Mantzoros C, Qu D, Lowell B, Maratos-Flier E, Flier JS: Role of leptin in the neuroendocrine response to fasting. Nature 1996;382:250–252.
88 Tartaglia LA, Dembski M, Weng X, Deng N, Culpepper J, Devos R, Richards GJ, Campfield LA, Clark FT, Deeds J, Muir C, Sanker S, Moriarty A, Moore KJ, Smutko JS, Mays GG, Woolf EA, Monroe CA, Tepper RI: Identification and expression cloning of a leptin receptor, OB-R. Cell 1995; 83:1263–1271.
89 Considine RV, Sinha MK, Heiman ML, Kriauciunas A, Stephens TW, Nyce MR, Ohannesian JP, Marco CC, McKee LJ, Bauer TL, Caro JF: Serum-immunoreactive leptin concentrations in normal-weight and obese humans. N Engl J Med 1996;334:292–295.
90 Montague CT, Farooqi IS, Whitehead JP, Soos MA, Rau H, Wareham NJ, Sewter CP, Digby JE, Mohammed SN, Hurst JA, Cheetham CH, Earley AR, Barnett AH, Prins JB, O'Rahilly S: Congenital leptin deficiency is associated with severe early-onset obesity in humans. Nature 1997;387:903–908.
91 Strobel A, Issad T, Camoin L, Ozata M, Strosberg AD: A leptin missense mutation associated with hypogonadism and morbid obesity. Nat Genet 1998;18:213–215.
92 Clément K, Vaisse C, Lahlou N, Cabrol S, Pelloux V, Cassuto D, Gourmelen M, Dina C, Chambaz J, Lacorte JM, Basdevant A, Bougneres P, Lebouc Y, Froguel P, Guy-Grand B: A mutation in the human leptin receptor gene causes obesity and pituitary dysfunction. Nature 1998;392:398–401.
93 Farooqi IS, Jebb SA, Langmack G, Lawrence E, Cheetham CH, Prentice AM, Hughes IA, McCamish MA, O'Rahilly S: Effects of recombinant leptin therapy in a child with congenital leptin deficiency. N Engl J Med 1999;341:879–884.
94 Mason AJ, Hayflick JS, Zoeller T, Young WS III, Phillips HS, Nikolics K, Seeburg PH: A deletion truncating the gonadotropin-releasing hormone gene is responsible for hypogonadism in the *hpg* mouse. Science 1986;234:1366–1371.
95 Nakayama Y, Wondisford FE, Lash RW, Bale AE, Weintraub BD, Cutler GB Jr, Radovick S: Analysis of gonadotropin-releasing hormone gene structure in families with familial central precocious puberty and idiopathic hypogonadotropin hypogonadism. J Clin Endocrinol Metab 1990;70: 1233–1238.
96 Weiss J, Adams E, Whitcomb RW, Crowley WF Jr, Jameson JL: Normal sequence of the gonadotropin-releasing hormone gene in patients with idiopathic hypogonadotropic hypogonadism. Biol Reprod 1991;45:743–747.
97 Schwanzel-Fukuda M, Pfaff DW: Origin of luteinizing hormone-releasing hormone neurons. Nature 1989;338:161–164.
98 Schwanzel-Fukuda M, Bick D, Pfaff DW: Luteinizing hormone-releasing hormone (LHRH)-expressing cells do not migrate normally in an inherited hypogonadal (Kallmann's) syndrome. Mol Brain Res 1989;6:311–326.

François P. Pralong, MD, Division of Endocrinology, BH 19-707, University Hospital
CH–1011 Lausanne (Switzerland)
Tel. +41 21 314 0619, Fax +41 21 314 0630, E-Mail francois.pralong@chuv.hospvd.ch

Rappaport R, Amselem S (eds): Hypothalamic-Pituitary Development. Genetic and Clinical Aspects.
Endocr Dev. Basel, Karger, 2001, vol 4, pp 140–161

Molecular Pathology of Corticotroph Function

Adrian J.L. Clark, Francesca M. Swords

Department of Endocrinology, St Bartholomew's and the Royal London School of Medicine and Dentistry, London, UK

Introduction

Pathological defects of the adrenocorticotropin (ACTH)-secreting corticotroph cells of the anterior pituitary are uncommon. Both underactivity, in the form of isolated ACTH deficiency (OMIM *201400), and overactivity, in the form of corticotroph hyperplasia leading to Cushing's disease may occasionally be encountered in paediatric endocrinology practice. More widespread pituitary defects that include corticotroph cells – e.g. hypopituitarism – will not be discussed here. In this chapter we shall review what is currently known about the molecular physiology of the control and secretion of ACTH, before considering the molecular defects of this process as illustrated by both animal models and human genetic disease.

Hypothalamic Control of the Corticotroph

About 10% of the cells in the human anterior pituitary are capable of secreting ACTH, the 39-amino-acid peptide that is the principal regulator of glucocorticoid production by the adrenal gland. Now classical studies by Porter and Jones [1] demonstrated that substances in the hypophyseal-portal plasma resulted in ACTH-dependent adrenal cell stimulation measured as ascorbic acid release. Moreover, an intact pituitary gland was needed for this effect which was not seen when carotid plasma was used in place of hypophyseal-portal plasma. Since this

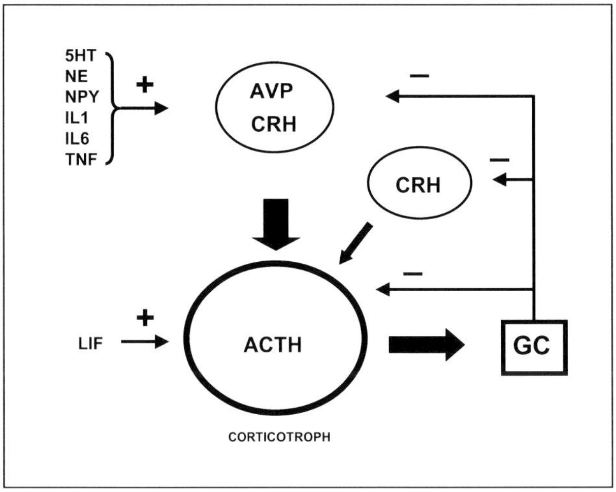

Fig. 1. Factors involved in the control of ACTH secretion. Both mixed CRH/AVP neurones and CRH neurones secrete their peptide products via the median eminence into the pituitary portal circulation to influence POMC gene expression, ACTH synthesis and secretion by the corticotroph cells of the anterior pituitary. ACTH stimulates glucocorticoid (GC) production which exerts a potent inhibitory feedback effect on both the hypothalamic nuclei and the corticotroph cell. Hypothalamic peptide secretion is stimulated by serotonin (5HT), noradrenaline (NE), neuropeptide Y (NPY), interleukins 1 (IL1) and 6 (IL6) and tumour necrosis factor (TNF). Leukaemia inhibitory factor (LIF) has a direct stimulatory effect on the corticotroph.

time it has become apparent that several factors can influence ACTH secretion, and these are summarized in figure 1. Two hypothalamic peptides in particular, corticotropin-releasing hormone and vasopressin, have a major role.

Corticotropin-Releasing Hormone (CRH)

CRH was first purified from sheep hypothalami in 1981, revealing it to be a 41-residue peptide with potent ACTH-releasing effects [2]. Cloning of the CRH gene followed in 1983 [3], and demonstrated that CRH is made as part of a large precursor from which the 41-residue biologically active peptide is derived by proteolytic cleavage (fig. 2a).

CRH is partially homologous to two non-mammalian peptides, sauvagine, obtained from frog skin, and urotensin I, found in fish [4]. A further mammalian CRH-like peptide, urocortin, was discovered in 1995 [5] and the human gene encoding urocortin was cloned soon afterwards [6]. The sequences of these peptides are compared in figure 2b.

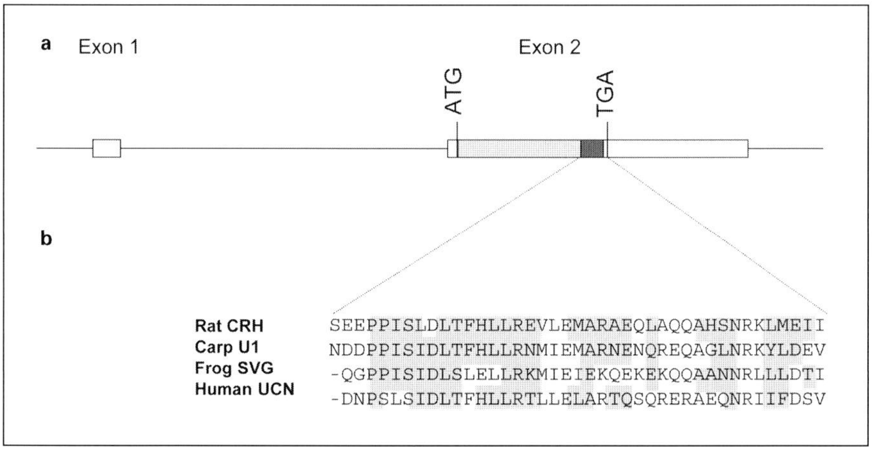

Fig. 2. ***a*** Structure of the CRH gene. Exon 1 encodes the RNA leader sequence, while exon 2 encodes the full-length 187 residue precursor peptide (pale grey shading). The 41 residues of the mature CRH peptide are contained in the C-terminal portion of this precursor (dark grey shading). ATG, initiator methionine codon, TGA, translation stop codon. ***b*** Comparison of CRH-like peptides including rat CRH, carp urotensin 1 (U1), frog sauvagine (SVG) and human urocortin (UCN). Regions of amino acid identity with rat CRH are shaded.

Localization studies performed primarily in the rat show that hypothalamic CRH neurones are located mainly in the parvocellular part of the paraventricular nucleus and project into the external zone of the median eminence [7, 8]. CRH neurones are also found in several other areas of the brain including the basal hypothalamus, brainstem, and cerebral cortex [8]. Manoeuvres such as adrenalectomy, volume depletion, hypoglycaemia and intracerebroventricular noradrenaline administration will increase CRH secretion into the hypophyseal portal circulation [9–12]. Glucocorticoid administration reverses this effect [12].

CRH Receptors

CRH was shown in 1982 to bind to specific sites in the pituitary and to stimulate adenylate cyclase and increased intracellular calcium accumulation [13–15]. A human pituitary CRH receptor, now known as CRH-R1, was cloned in 1993 and was revealed to be a member of the class B, glucagon/VIP/calcitonin receptor family of G-protein-coupled receptors [16–18]. This receptor binds CRH, sauvagine and urotensin I with equally high affinity and is highly expressed in the pituitary and many areas of the brain [19]. However, certain discrepancies between the distribution of CRH-R1 mRNA and CRH binding provided the clue that other CRH-Rs might exist, and the CRH-R2 was cloned in 1995 from rat brain and mouse heart [20–23]. This related receptor exhibits significantly greater affinity for sauvagine, urotensin and urocortin than for CRH [19]. CRH-R2 is expressed at only low levels in the pituitary [24].

Vasopressin

The nonapeptide arginine vasopressin (AVP) was first proposed to be a corticotrophin-releasing factor in 1955, although its effect in the rat and in humans is significantly weaker than that of CRH. The opposite is true in sheep where AVP is the principal regulator of ACTH secretion [25]. It was shown several years ago that there was a synergy between CRH and AVP in releasing ACTH in the rat [26], and it may be that this is central to the role of AVP in most species.

AVP is part of a large precursor, pre-pro-AVP-neurophysin, located on human chromosome 20p13 in close proximity to the very similar pre-pro-oxytocin-neurophysin gene. The two mammalian genes presumably arose by a gene duplication event at some time. The precursor is cleaved into AVP, neurophysin and the C-terminal glycopeptide, and neurophysin has a role as an AVP carrier protein. In the hypothalamus, two types of AVP neurone are found – those in the paraventricular and supraoptic nuclei with nerve terminals in the posterior pituitary, and those in the paraventricular nucleus with terminals in the median eminence. The former are responsible for the classical antidiuretic actions on water balance, while the latter population is involved in corticotroph regulation. This population co-localizes with CRH neurones, and both AVP and CRH can be demonstrated in the same secretory vesicles, although AVP-negative CRH-positive neurones are also located in this region [27]. Manipulations such as hypoglycaemia or hypotension that stimulate hypothalamic CRH secretion do not appear to affect AVP secretion [9, 10], although glucocorticoids have been shown to inhibit the quantities of AVP in the hypophyseal-portal circulation [28, 29].

Vasopressin Receptors

Three distinct receptors for AVP have been identified and cloned and all fall into the rhodopsin family (class A) of the G-protein-coupled receptor superfamily. The V1a receptor is expressed relatively widely, notably in vascular smooth muscle, and acts by activation of phospholipase C. The V2 receptor is located principally in the renal tubule where it regulates water resorption and is coupled to adenylate cyclase. Gene defects in the V2 receptor lead to X-linked nephrogenic diabetes insipidus [30]. The V1b receptor, sometimes called the V3 receptor, is a distinct gene product from the V1a receptor, and is expressed in the anterior pituitary. It is this receptor that is responsible for the actions of AVP on ACTH secretion. V1b expression is inhibited by adrenalectomy and stimulated by glucocorticoids in the rat [31]. This receptor was first cloned in 1994 [32], and is expressed in the pituitary, as well as in several regions of the brain and extrapituitary sites [33]. The V1b receptor is coupled to phospholipase C leading to an increase of intracellular calcium in the corticotroph and activation of protein kinase C [34–36]. This latter effect appears to be responsible for stimulating the cAMP response to CRH [37, 38], and thus provides one mechanism for producing synergy between the two factors.

Other Factors That Influence the Corticotroph

Noradrenaline, acting via a central α_1-adrenoreceptor pathway can regulate the waking and post-prandial release of ACTH in the human [39–41]. This effect is almost certainly mediated by the effect of increased cAMP stimulation on CRH synthesis in the paraventricular nucleus of the hypothalamus since injection of noradrenaline into this region in the rat has been shown to stimulate CRH mRNA production [42]. There may also be an additional effect of chronic noradrenergic stimulation in inhibiting expression of the glucocorticoid receptor mRNA at this site, thereby reducing the glucocorticoid repression of CRH synthesis and release [43].

Neuropeptide Y, a 36-residue peptide that is widely distributed in the brain, also acts in a similar manner to noradrenaline in stimulating CRH synthesis and release in the hypothalamus [44].

5-Hydroxytryptamine: Serotonergic agonists, including drugs that stimulate 5-hydroxytryptamine (5-HT) release, can stimulate increased gene expression of CRH and release into the hypophyseal portal circulation [45] as well as having direct actions on ACTH release by the pituitary [46], possibly by a potentiation of the actions of CRH and AVP. Actions of 5-HT appear to be mediated primarily by 5-HT1A, 5-HT2A and 5-HT2C receptors [47].

Cytokines: Leukocyte inhibitory factor (LIF), interleukins 1 and 6 and tumour necrosis factor (TNF) can all be shown to stimulate ACTH secretion. While the interleukins appear to act by stimulating CRH gene expression and secretion, LIF has been shown to act on LIF receptors on the corticotroph resulting in stimulation of proopiomelanocortin gene expression and ACTH release directly [48].

Proopiomelanocortin

The Proopiomelanocortin Gene (POMC)

POMC has now been studied in a wide variety of species ranging from fish to the human. Most of the studies of its tissue-specific and hormonal regulation however have been performed in the rat and human. The genes from these two species are similar and much information is interchangeable between them, although

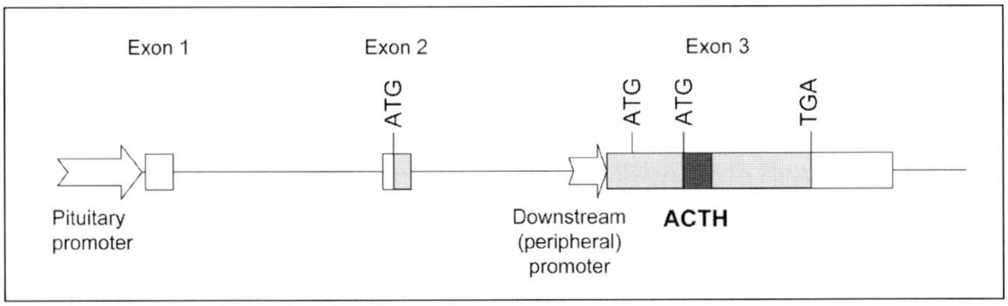

Fig. 3. Structure of the POMC gene. Exon 1 encodes the RNA leader sequence, exon 2 encodes the initiator methionine (ATG), the signal peptide and several N-terminal residues of the precursor peptide, the remainder of which is encoded by exon 3. Corticotroph expression is determined by the upstream 'pituitary' promoter (longer white arrowhead), whereas peripheral expression of the short POMC mRNA is determined by the 'downstream' promoter (shorter white arrowhead). Translation of these shorter transcripts initiates from the initiator methionines (ATG) indicated in exon 3. The precursor peptide coding region is shaded light grey and the ACTH coding region is dark grey.

there are specific differences. In the subsequent discussion, the data reviewed relates to the human gene unless otherwise stated.

POMC is a single copy gene that is located on the short arm of human chromosome 2 (2p15) [49]. A second pseudogene has been identified in the mouse [50]. The full-length gene extends over approximately 12 kbp and consists of 3 exons of which exon 2 (152 bp) and 3 (833 bp) contain the coding regions. Exon 1 (87 bp) encodes the RNA leader sequence for this gene [51–53]. These principal architectural features are summarized in figure 3.

POMC is expressed as one of two types of RNA transcript which are largely determined by the tissue of origin. The corticotroph cells of the pituitary classically express *POMC* from a comparatively strong promoter located immediately upstream of exon 1, to produce an RNA transcript of approximately 1,200 nt [54, 55]. Only very weak activity of this promoter can normally be detected in other tissues, although this may refelect the low prevalence of cells expressing this 'full-length' transcript in some of those tissues. Of particular relevance in this respect is the activity in the arcuate nucleus of the brain from which α-MSH is produced and exerts a major role in the control of appetite.

Most extrapituitary tissues can express *POMC* from the very weak downstream promoter located at the beginning of exon 3 [54–57]. Notable sites of expression from this promoter include the testis, ovary and placenta [58, 59]. The translation product of these 'peripheral POMC' transcripts is derived entirely from exon 3 and therefore lacks the secretory signal sequence encoded in exon 2. As a result, the evidence shows that this peptide is not normally secreted [60].

The pituitary promoter has been shown by several authors to function as a promoter by its ability to drive expression of various reporter genes after transfection into cells. Although there remains some debate over certain aspects of *POMC* promoter function, a number of conclusions can be drawn. A key feature is the presence of a typical TATA box located 27 bp upstream of the transcription start site which positions the RNA polymerase II complex on the gene correctly.

However, a promoter including only this element is very weak and several other enhancer elements are required. These include an upstream regulatory element which was initially thought to be a typical CAAT-box element located at –62bp. More recent evidence suggests this is in fact the binding site for *Nur77/ NGFI-B* [61], an orphan nuclear receptor, which has been shown to bind to this response element either as a homodimer [62] or as a heterodimer with the related factor *Nurr1* [63]. Furthermore, CRH stimulation of corticotrophs stimulates expression of the *Nur77* gene in primary cultures of pituitary cells providing a mechanism for the positive regulation of POMC gene expression by CRH [61]. In contrast, glucocorticoid inhibition of *POMC* expression which has been recognised to be mediated through this region for many years [64] is mediated by a direct antagonism between ligand occupied glucocorticoid receptor and *Nur77* at the level of a protein-protein interaction [65, 66].

At least two other regions of the POMC promoter have been identified as being crucial in conferring the tissue specificity of expression. These are a site originally called the corticotroph upstream transcription element (CUTE), and a second more distal site which was shown to bind a homeobox protein, *Ptx1*, as a monomer [67].

Ptx1 is expressed early in development [68], and contributes to the formation of several cell lineages within the pituitary by means of synergy with other cell-specific factors [69–72]. In addition, *Ptx1* has a major role in hind limb patterning [73, 74]. Thus *Ptx1* contributes not only to corticotroph development, but also to somatotroph-lactotroph development with Pit1 and gonadotroph development with SF1. *Ptx1* is a member of the Drosophila *bicoid* homeobox family which also includes *Ptx2* and *3* and *Otx1* and *2*. *Ptx1* co-localizes to a region on chromosome 5 (5q31) close to that to which the Treacher Collins syndrome – a craniofacial maldevelopment syndrome – has been mapped. However, it is now clear that *Ptx1* is not causative in this syndrome. Mice in which *Ptx1* has been deleted exhibit defects of hind limb formation and maldevelopment of the anterior pituitary and cleft palate [75].

The CUTE element has been shown to bind a heterodimeric pair of helix-loop-helix (H-L-H) transcription factors which include the corticotroph-specific *NeuroD1* and one of several ubiquitous H-L-H factors such as *Pan1* [76, 77]. *NeuroD1* is also known as *beta2* and is involved in pancreatic development and differentiation. Indeed, mutations in this factor have been associated with maturity-onset diabetes of the young [78], and mice in which it has been deleted have atypical pancreatic islet development [79]. However, it is clear that the *NeuroD1* heterodimer interacting with the *POMC* gene synergizes with *Ptx1* to determine corticotroph-specific expression of *POMC*. New data is currently emerging on a further corticotroph-specific factor that interacts with *Ptx1* and *NeuroD1*.

POMC Processing

POMC provides the archetypal example of peptide hormone precursor processing. In now classical work the existence of a single precursor peptide containing both ACTH and β-lipotropin was originally demonstrated by three groups at about the same time [80–82]. These findings were confirmed with the cloning and sequencing of the bovine cDNA by Nakanishi et al. [83]. This revealed the repeat-like nature of the precursor which contained three MSH sequences (α-, β- and γ-MSH) as well as β-endorphin. Distinct peptides are separated by dibasic amino acid cleavage sites.

The full-length precursor is synthesized with a short (26-residue) signal peptide sequence needed for its passage across the endoplasmic reticulum and ultimately secretion of peptide products. This signal peptide is cleaved from POMC at, or shortly after, passage across the membrane. Two disulphide bridges are formed between cysteines 2 and 24 and cysteines 8 and 20 at its N-terminal end [84]. The resulting 8–20 loop has been shown to be the element that binds to membrane-bound carboxypeptidase E in the trans-Golgi and which acts as a sorting receptor directing bound polypeptides into the regulated secretory pathway [85]. In this pathway the precursor is subsequently glycosylated on asparagine 65 and 136 (in the γ-MSH and CLIP sequence respectively [84]) as well as O-linked glycosylation on serine 142 [86]. Serine 142 is phosphorylated. These processing events are summarized in figure 4.

Cleavage of the POMC precursor into biologically functional peptides is a critical process that occurs in the secretory granule. The main enzymes involved in this process are the prohormone convertases, soluble subtilsin-like peptidases which are secreted into the secretory vesicle. In the anterior pituitary corticotroph, prohormone convertase 1 (PC1, also sometimes known as PC3) cleaves the molecule as shown in figure 5 to generate N-POC and β-lipotropin initially, and then cleaving N-POC to form pro-γ-MSH, joining peptide and ACTH. In the rodent neurointermediate lobe and in the few neurointermediate cells in the human pituitary, a second prohormone convertase, PC2, is expressed in addition to PC1 and this can cleave ACTH into α-MSH and corticotropin-like intermediate lobe peptide (CLIP) and β-lipotropin into β-endorphin and γ-lipotropin [84, 87, 88].

The action of PC2 appears to require a second polypeptide, 7B2, as a cofactor [89], and mice in which *7B2* has been deleted show no PC2 activity, but

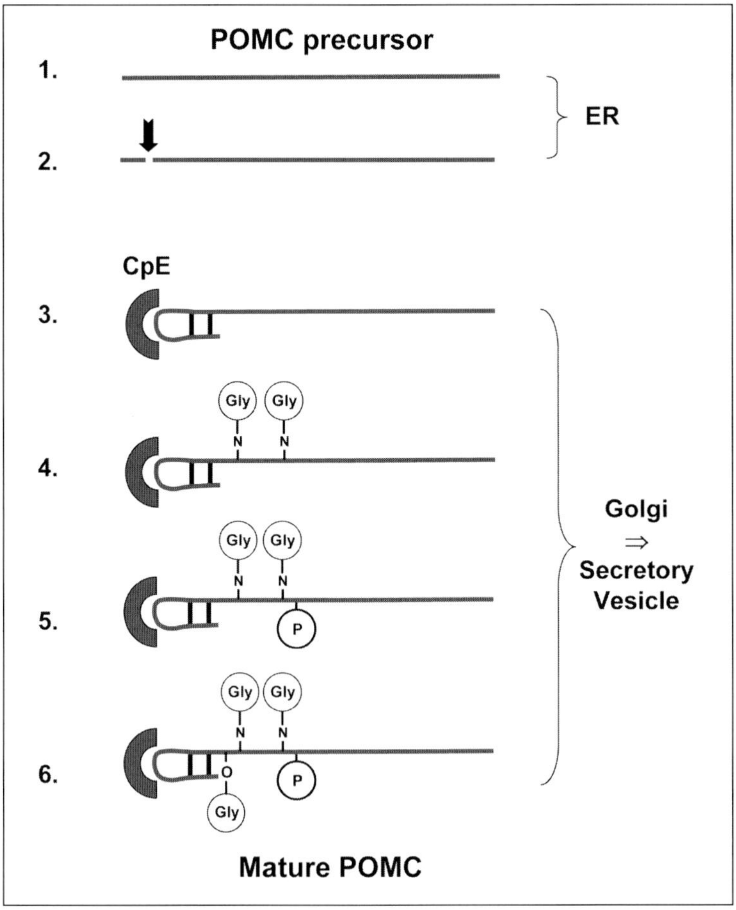

Fig. 4. The full-length POMC translation product (1) is initially cleaved co-translationally in the endoplasmic reticulum to remove the signal peptide (2). A number of other steps then take place in the Golgi apparatus. Disulphide bonds are formed between Cys 2 and Cys 26, and Cys 8 and Cys 20 (3). The 8–20 loop acts as a binding target for membrane associated carboxypeptidase E (CpE) which sorts the peptide into the regulated secretory pathway (3). N-linked glycosylation on positions 65 and 136 (4), phosphorylation on serine 142 (5) and O-linked glycosylation on threonine 45 (6) also occurs.

develop neurointermediate lobe Cushing's disease [90]. Several other enzymes are involved in POMC processing. Carboxypeptidase H (carboxypeptidase E) removes residual basic amino acids at cleavage sites. Peptidylglycine α-amidating monooxygenase (PAM) amidates the C-terminal tail of glycine-extended MSH peptides [91].

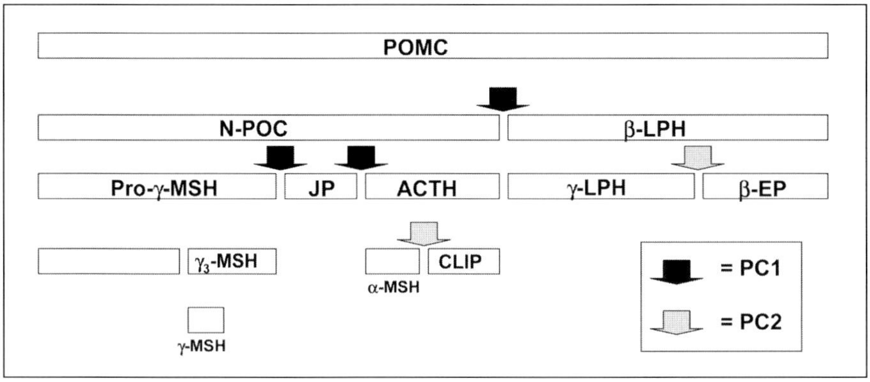

Fig. 5. Processing and cleavage of POMC. The mature POMC precursor peptide is sequentially cleaved by PC1 in the anterior pituitary corticotroph. In neurointermediate lobe and other cell types, cleavage by PC2 follows to release β-MSH and/or β-endorphin. Carboxypeptidase H (not shown) removes residual basic amino acids at cleavage sites. β-LPH, β-lipotropin; JP, joining peptide; γ-LPH, γ-lipotropin; CLIP, corticotropin-like intermediate lobe peptide.

Pathological Defects of Corticotroph Function

As indicated in the Introduction, molecular defects of corticotroph function in clinical practice are rare. The pathophysiology of corticotroph tumours is far from clear and the topic is too extensive to be adequately considered here. This has recently been reviewed in some detail elsewhere [92].

Isolated ACTH deficiency is the most probable outcome of a failure of the processes reviewed above, but in many of these cases the aetiology is probably autoimmune often in the form of lymphocytic hypophysitis [93]. It has been suggested that the primary antigen in these cases is a product of the corticotroph other than a POMC-derived peptide [94]. There is undoubtedly heterogeneity amongst those cases that appear to be of genetic origin, although the defect appears to lie at the level of the pituitary in the majority in that there is frequently a failure of ACTH to respond to CRH. Possible targets for such a defect clearly include CRH and AVP, their cognate receptors in the pituitary, POMC and the various factors involved in both its expression and processing in the corticotroph to produce ACTH. These factors are summarised in table 1.

CRH and CRH-R Defects

An inactivating mutation of CRH or CRH-R1, the pituitary CRH receptor, would provide a plausible candidate for isolated ACTH deficiency. CRH gene

Table 1. Summary of genetic defects with an established or probable influence on corticotroph function

Candidate gene	Human chromosome	Animal model	Human disease
Ptx1	5q31	Maldevelopment of anterior pituitary Cleft palate Defective hind limb formation [75]	Treacher-Collins syndrome – excluded
Neuro D	2q32	Atypical pancreatic islet development [79]	Maturity-onset diabetes of the young [78]
CRH	8q13	CRH-responsive isolated ACTH deficiency [95]	Not described
AVP	20p13	Brattleboro rat – diabetes insipidus [105, 106]	Autosomal dominant diabetes insipidus [107]
CRH-R1	17q12–22	↓ ACTH response to stress Behavioural defects [102, 103]	Not described
V1b-R	1q32	Not described	Not described
POMC	2p15	Isolated ACTH deficiency Hyperphagia and obesity Yellow hair [114]	Isolated ACTH deficiency Severe obesity Red hair [108, 112, 113]
CPE	4	Obesity Insulin-dependent diabetes [85, 115]	Not described
PC1	5q15	Not described	Isolated ACTH deficiency Severe obesity Impaired glucose tolerance Hypogonadotropic hypogonadism [116, 117]
PC2	20p11.2	Not described	Not described
7B2	15q11–15	Neurointermediate lobe Cushing's disease [90]	Not described

targeting in mice results in animals that have markedly defective ACTH and corticosterone values basally and in response to a variety of stresses. Interestingly, corticotroph cells develop normally and contain immunostainable ACTH, yet ACTH deficiency is sufficiently marked as to result in failure of development of the adrenal zona fasciculata. These mice are completely viable and survive normally. However, homozygous CRH^-/CRH^- mice born to similarly homozygous CRH^-/CRH^- mothers fail to survive as a result of failed lung maturation. This situation can be reversed by glucocorticoid administration in pregnancy [95, 96]. The behavioural abnormalities seen in CRH-R1-deficient mice (see below) are not apparent in these CRH knockout animals [97] implying that urocortin or an as yet unidentified CRH-like molecule is responsible for these actions. These observations argue in favour of the case for CRH gene defects as candidates for isolated ACTH deficiency. However, many patients with isolated ACTH deficiency fail to respond to CRH [e.g. 98] arguably making CRH defects per se less probable, whereas in contrast, administration of CRH to CRH gene-deleted mice restores ACTH secretion and rhythmicity [99]. No human subject with a CRH gene defect has yet been identified.

In some cases of isolated ACTH deficiency a response to lysine vasopressin can be demonstrated in patients with a failed response to CRH [100], supporting the candidacy of the CRH-R. Kyllo et al. [101] reported that the CRH-R1 locus on chromosome 8 co-segregated with isolated ACTH deficiency in 2 affected siblings. However, no reports of mutations in this receptor have appeared from this, or any other group, since this publication.

Evidence *against* the candidacy of the CRH-R1 in isolated ACTH deficiency comes from the study of mice in which this gene has been deleted. Such mice have been generated by two groups with broadly similar findings [102, 103]. These show that basal ACTH concentrations are similar to those of wild-type mice, but that it is the response to stress which reveals the defect in these animals. The maintenance of basal ACTH is largely due to an enhancement of the role of AVP in that AVP antiserum markedly reduced resting ACTH values in these animals [103]. Furthermore, behavioural defects in the form of impaired anxiety-like and cognitive activity were observed [104]. Thus, to the extent that such animals can provide a model for human disease, the case for CRH-R1 defects in the pathogenesis of classical isolated ACTH deficiency is not strengthened.

Vasopressin and the V1b Receptor

The Brattleboro rat provides an intensively studied animal model of vasopressin deficiency. However, there is evidence of only relatively subtle defects of the pituitary-adrenal axis in this model, and animals are not ACTH or corticosterone deficient [105, 106]. Vasopressin gene defects occur in humans and give rise to autosomal dominant diabetes insipidus. As with the animal model, these patients do not have ACTH deficiency [107]. It would seem inevitable that the antidiuretic actions of AVP would be a far more dominant phenomenon, and it is therefore reasonable to exclude this gene as a candidate for isolated ACTH deficiency.

At the present time no animal model in which the V1b receptor has been deleted has been described, and such defects have not been reported in humans.

The POMC Gene

Defects in the POMC gene encoding ACTH would appear to provide the obvious candidate for investigation in isolated ACTH deficiency. However, as will be seen, typical isolated ACTH deficiency in the absence of any other phenotype has not yet been linked with *POMC*. Furthermore, mouse models of POMC deficiency have been relatively slow to be developed, with the result that this syndrome was first described in an unrelated pair of children [108]. Both children had had isolated ACTH deficiency from a young age in addition to severe early-onset obesity and the presence of red hair. One case was a compound heterozygote for two different mutations of *POMC*. One mutation was a frameshift at codon 144 (7133 del C) leading to a peptide product with a short nonsense sequence after the sixth residue of the α-MSH sequence of ACTH, and the other was a single base change leading to a premature stop codon at position 104 (7013 G → T), also leading to a peptide truncated prior to the ACTH sequence. The same compound heterozygous mutation was found in the proband's brother who had died of hepatic failure with undiagnosed adrenal insufficiency.

The second patient was homozygous for a point mutation in the 5′ untranslated region of exon 2 (3804 C → A) which creates an additional (out-of-frame) start codon. This appears to abolish translation of *POMC* with translation of an unrelated protein in an alternative reading frame [108].

In both patients the major clinical presentation was that of glucocorticoid deficiency. Both were confirmed to lack ACTH in the plasma, and there was absent ACTH staining in histological sections of the pituitary of the infant who died of hepatic failure. Both patients have been symptomatically well since the institution of hydrocortisone therapy with no further episodes of hypoglycaemia or cholestasis. However, both children have developed severe obesity within the first year of life. All 3 children were noted to have pale skin with red-orange hair, and both living children have undetectable α-MSH levels. The parents in both families have been confirmed to be heterozygous for the *POMC* mutations but interestingly none are affected by hypocortisolaemia, obesity or red hair.

The inclusion of red hair and obesity in the phenotype is the consequence of the important role of α-MSH as an agonist for the MC1 receptor in causing pigment deposition in the hair follicle, and as an agonist for the MC4 receptor in the hypothalamus where it has a vital role in the control of appetite and food intake. Defects in the MC4 receptor itself give a similar syndrome of severe early-onset obesity without ACTH deficiency or hair pigment changes in both mice and humans [109–111].

Since the publication of these first 2 cases of human POMC deficiency, the syndrome has been recognised in additional cases. Krude and Gruters [112] reported 2 additional cases: a male infant from Slovenia also with isolated ACTH deficiency, red hair and obesity who had compound heterozygous mutations leading to truncation of the POMC gene product (6851A → T, 6996 del G); another male infant from the Netherlands was homozygous for the same point mutation in the 5′ untranslated region of *POMC* as reported previously (3804C → A). A third patient from Japan with this phenotype is homozygous for a deletion-insertion mutation in exon 2 leading to termination upstream of ACTH (3895 del GTG ins AC) [113]. These mutations are summarized in figure 6.

A murine model for *POMC* deficiency has now been developed using gene-targeting techniques. In this model, the entire third exon of *POMC* has been

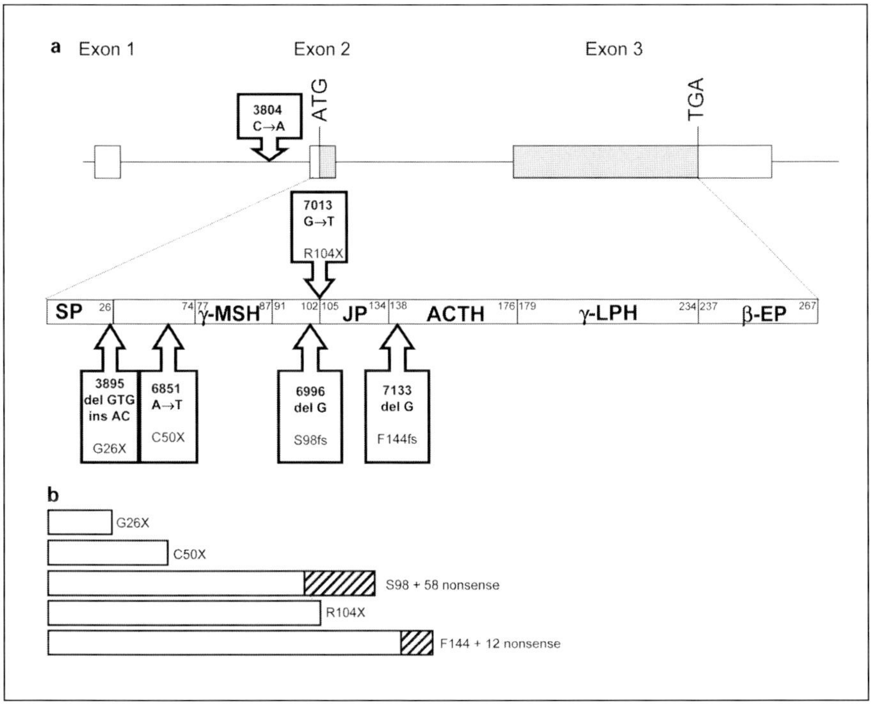

Fig. 6. POMC mutations in humans. ***a*** The structure of the POMC gene and peptide precursor based on figures 3 and 4b indicating the location of the mutations of human POMC (in arrowed boxes) described to date. DNA mutations are in bold, amino acid consequences of the mutation using the single letter amino acid code are in normal font. Numbers in boxes representing peptide indicate sites of cleavages. Fs, frameshift; SP, signal peptide. Numbering of nucleotide positions is as in [108, 112, 113]. Numbering of amino acids is as in [51–53]. ***b*** POMC peptide products resulting from these mutations including, when present, the length of nonsense sequence resulting from the frameshift shown in the hatched box.

replaced with a neomycin resistance cassette. It is interesting to note that the frequency of homozygosity in live born pups is only one quarter of that predicted, suggesting that loss of embryonic POMC-derived peptides has a deleterious effect on foetal survival and development. These mice had severe glucocorticoid deficiency with virtually absent adrenal development, a yellowish pigmentation, and obesity resulting from greater food intake and greater fat deposition. Administration of a stable α-MSH analogue to these animals resulted in a reversal of weight gain and a loss of yellow pigmentation [114].

Defects in POMC Processing

Many of the steps involved in processing the POMC molecule outlined above and in figures 4 and 5 are common to the processing of many other peptides and proteins, and thus defects in these processes may have more severe consequences than ACTH deficiency alone.

One of the first steps in POMC processing is the association of the polypeptide with carboxypeptidase E and sorting into the regulated secretory pathway. A mouse model for a defect at this level exists in the form of the *Cpe^fat^* mouse, a naturally occurring mutant that has a point mutation in the gene encoding carboxypeptidase E that renders it unstable [85, 115].

In these mice the POMC gene product is incorrectly sorted to the constitutive secretory pathway and secreted in an unregulated manner without specific post-translational modification. The *Cpe^fat^* mouse demonstrates elevated plasma levels of POMC precursor which cannot be inhibited by dopamine. Furthermore, levels of α-MSH are reduced with increased levels of larger ACTH precursors. It is likely that some PC1 processing takes place because this enzyme activity occurs earlier in the secretory pathway including some activity at the Golgi, whereas PC2 activity, required for formation of α-MSH occurs later in mature secretory vesicles. The *Cpe^fat^* mouse is markedly obese as a result of α-MSH deficiency, and has insulin-dependent diabetes as a result of impaired cleavage of insulin from proinsulin. Corticosterone deficiency is not apparent.

No human equivalent of this defect has yet been recognised, and is probably more likely to present with early-onset diabetes and obesity than with ACTH deficiency. However, a defect in PC1 has been described only in the human and an equivalent mouse model is not yet available.

This female patient exhibited extreme childhood obesity and hypocortisolaemia resulting from ACTH deficiency. However, in addition she demonstrated abnormal glucose homeostasis and hypogonadotrophic hypogonadism. Circulating POMC (ACTH precursor) concentrations were elevated as was plasma proinsulin with very low plasma insulin levels. This combination of abnormalities is again highly suggestive of defective processing [116].

A compound heterozygous mutation in PC1 was found. One missense mutation (G483R) gave rise to impairment of processing of pro-PC1 itself leading to its retention in the endoplasmic reticulum. The second mutation is an A for C substitution in the intron donor splice site which causes skipping of exon 5 with the loss of 26 residues, a frameshift and creation of a premature stop codon within the catalytic domain of PC1. Thus this patient is PC1 deficient which provides a plausible explanation for her ACTH deficiency, obesity and insulin-dependent diabetes [117].

A second patient with a syndrome of isolated ACTH deficiency and severe obesity has also been described although the underlying molecular defect has not been defined. This patient presented with hypoglycaemia at 11 h of age. No other clinical features were noted until the age of 2 years after which the patient developed hyperphagia, and despite restricted access to food, she has now developed marked obesity. In contrast to patients with POMC deficiency, she does not have red hair [118].

On biochemical investigation, the patient was found to have undetectable cortisol levels during stress (hypoglycaemia) and during a short Synacthen test, with a suboptimal rise after a depot Synacthen test, suggesting the presence of functional adrenocortical tissue. She had undetectable ACTH levels which did not rise with CRH stimulation suggesting that the defect was at the level of the corticotroph. However, plasma levels of N-POC and POMC were normal or elevated. Both rose appropriately with CRH and were suppressed appropriately with dexamethasone. Further evaluation revealed normal levels of β-endorphin. The patient had a normal ratio of insulin:proinsulin, suggesting the presence of functional PC1 and CPE. The responses of glucagon to glucose and glucagon-like peptide1 to a test meal were also normal [118].

The POMC gene of this patient was normal on DNA sequencing of exons 2 and 3 suggesting that the defect in this case lay not in the genomic DNA encoding ACTH or the cleavage region for ACTH and β-LPH, but rather in the processing of this gene. However, all four exons, the exon/intron boundaries, and the promoter region of PC1 have been sequenced and are normal [119]. A defect in PC2 is unlikely since ACTH should be produced normally if this were the case. Thus while a specific defect in POMC processing seems highly likely, its nature remains elusive, reflecting the gaps in our current understanding of these events.

Conclusion

In this chapter we have reviewed the molecular physiology and pathophysiology of the corticotroph cell. Disease of the corticotroph takes the form of hyperplastic/neoplastic lesions (Cushing's disease) and corticotroph failure (isolated ACTH deficiency). Proliferative defects of the corticotroph are not well understood and probably relate more to proliferative defects in general than to corticotroph-specific lesions, and thus these have not been discussed. The reader is referred to other major reviews on this subject [e.g. 92].

The genes that could potentially give rise to isolated ACTH deficiency are summarized in table 1. These human diseases are all very uncommon, and none of the identified genetic defects give rise to 'typical' isolated ACTH deficiency. More frequently they cause combined syndromes which include childhood obesity as a result of α-MSH deficiency. This probably reflects our relatively naive understanding of the mechanisms of ACTH secretion, although the possibility of

CRH gene defects has probably not been explored to its fullest extent. Undoubtedly, technological progress in DNA sequencing and mutation detection will facilitate further discoveries in this field in the coming years.

References

1 Porter JC, Jones JC: Effect of plasma from hypophyseal-portal vessel blood on adrenal ascorbic acid. Endocrinology 1956;58:62–67.

2 Vale W, Speiss J, Rivier C, Rivier J: Characterization of a 41-residue ovine hypothalamic peptide that stimulates secretion of corticotropin and beta-endorphin. Science 1981;213:1394–1397.

3 Shibahara S, Morimoto Y, Furutani Y, Notake M, Takahashi H, Shimizu S, Horikawa S, Numa S: Isolation and sequence analysis of the human corticotropin-releasing factor precursor gene. EMBO J 1983;2:775–779.

4 Lovejoy DA, Balment RJ: Evolution and physiology of the corticotropin-releasing factor family of neuropeptides in vertebrates. Gen Comp Endocrinol 1999;115:1–22.

5 Vaughan J, Donaldson C, Rittencourt J, Porrin MH, Lewis K, Sutton S, Chan R, Turnbull AV, Lovejoy D, Rivier C, Rivier J, Sawchenko PE, Vale W: The mammalian neuropeptide urocortin, related to fish urotensin I and to corticotropin-releasing factor. Nature 1995;378:287–292.

6 Donaldson CJ, Sutton SW, Perrin MH, Corrigan AZ, Lewis KA, Rivier JE, Vaughan JM, Vale WW: Cloning and characterization of human urocortin. Endocrinology 1996;137:2167–2170.

7 Bloom FE, Battenberg ELF, Rivier J, Vale W: Corticotrophin-releasing factor: Immunoreactive neurones and fibres in rat hypothalamus. Regul Pept 1982;4:43–48.

8 Swanson LW, Sawchenko PE, Rivier J, Vale WW: Organization of corticotropin-releasing factor immunoreactive cells and fibres in the rat brain: An immunohistochemical study. Neuroendocrinology 1983;36:165–186.

9 Plotsky PM, Bruhn TO, Vale W: Evidence for multifactor regulation of the adrenocorticotropin secretory response to hemodynamic stimuli. Endocrinology 1985;116:633–639.

10 Plotsky PM, Bruhn TO, Vale W: Hypophysiotropic regulation of adrenocorticotropin secretion in response to insulin-induced hypoglycemia. Endocrinology 1985;117:323–329.

11 Plotsky PM: Facilitation of immunoreactive corticotrophin-releasing factor secretion into the hypophysial-portal circulation after activation of catecholaminergic pathways or central norepinephrine injection. Endocrinology 1987;121:924–930.

12 Fink G, Robinson ICAF, Tannahill LA: Effects of adrenalectomy and glucocorticoids on the peptides, CRF-41, AVP and oxytocin in rat hypophysial portal blood. J Physiol 1988;401:329–345.

13 Labrie F, Veilleux R, Lefevre R, Coy DH, Sueiras-Diaz J, Schally AV: Corticotropin-releasing factor stimulates accumulation of adenosine 3′,5′-monophosphate in rat pituitary corticotropes. Science 1982;216:1007–1008.

14 Aguilera G, Harwood JP, Wilson JX, Morell J, Brown JH, Catt KJ: Mechanisms of action of corticotropin-releasing factor and other regulators of corticotropin release in rat pituitary cells. J Biol Chem 1983;258:8039–8045.

15 Abou-Samra AB, Catt J, Aguilera G: Calcium-dependent control of corticotropin release in rat anterior pituitary cell cultures. Endocrinology 1987;121:965–971.

16 Chang CP, Pearse RV, O'Connell S, Rosenfeld MG: Identification of a seven transmembrane helix receptor for corticotropin-releasing factor and sauvagine in mammalian brain. Neurone 1993;11: 1187–1195.

17 Chen R, Lewis KA, Perrin MH, Vale WW: Expression cloning of a human corticotropin-releasing factor receptor. Proc Natl Acad Sci USA 1993;90:8967–8971.

18 Vita N, Laurent P, Lefort S, Chalon P, Lelias JM, Kaghad M, LeFur G, Caput D, Ferrara P: Primary structure and functional expression of mouse pituitary and human brain corticotropin-releasing factor receptors. FEBS Lett 1993;335:1–5.

19 Speiss J, Dautzenberg FM, Sydow S, Hauger RL, Ruhmann A, Blank T, Radulovic J: Molecular properties of the CRF receptor. Trends Endocrinol Metab 1998;9:140–145.
20 Lovenberg TW, Liaw CW, Grigoriadis DE, Clevenger W, Chalmers DT, DeSouza EB, Oltersdorf T: Cloning and characterization of a functionally distinct corticotropin-releasing hormone receptor subtype from rat brain. Proc Natl Acad Sci USA 1995;92:836–840.
21 Kishimoto T, Pearse RV II, Lin CR, Rosenfeld MG: A sauvaging/corticotropin-releasing factor receptor expressed in heart and skeletal muscle. Proc Natl Acad Sci USA 1995;92:2969–2973.
22 Perrin MH, Donaldson C, Chen R, Blount A, Berggren T, Bilezikjian L, Sawchewnko PE, Vale W: Identification of a second corticotropin-releasing factor receptor gene and characterization of a cDNA expressed in heart. Proc Natl Acad Sci USA 1995;92:2969–2973.
23 Stenzel P, Kesterson R, Yeung W, Cone RD, Rittenberg MB, Stenzel-Poore MP: Identification of a novel murine receptor for corticotrophin-releasing hormone expressed in the heart. Mol Endocrinol 1995;9:637–645.
24 Chalmers DT, Lovenberg TW, De Souza EB: Localization of novel corticotrophin-releasing factor receptor (CRH_2) mRNA expression to specific subcortical nuclei in rat brain: Comparison with CRF_1 receptor mRNA expression. J Neurosci 1995;15:6340–6350.
25 Van der Pavert SA, Clarke IJ, Rao A, Vrana KE, Schwartz J: Effects of vasopressin and elimination of corticotrophin-releasing hormone target cells on pro-opiomelanocortin mRNA and adrenocorticotrophin secretion in ovine anterior pituitary cells. J Endocrinol 1997;154:139–147.
26 Gilles GE, Linton EA, Lowry PJ: Corticotropin-releasing activity of the new CRF is potentiated several times by vasopressin. Nature 1982;299:355–357.
27 Whitnall MH, Mezey E, Gainer H: Co-localization of corticotropin-releasing factor and vasopressin in median eminence neurosecretory vesicles. Nature 1985;317:248–250.
28 Antoni FA, Fink G, Sheward WJ: Corticotrophin-releasing peptides in rat hypophysial portal blood after paraventricular lesions: A marked reduction in the concentration of corticotrophin-releasing factor-41, but no change in vasopressin. J Endocrinol 1990;125:175–183.
29 Sapolsky RM, Armanini MP, Packan DR, Sutton SW, Plotsky PM: Glucocorticoid feedback inhibition of adrenocorticotrophic hormone secretagogue release. Relationship to corticosteroid receptor occupancy in various limbic sites. Neuroendocrinology 1990;51:328–336.
30 Van den Ouweland AM, Dreesen JCFM, Verdijk M, Knoers NVAM, Monnens LAH, Rocchi M, van Oost BA: Mutations in the vasopressin type 2 receptor gene (AVPR2) associated with nephrogenic diabetes insipidus. Nat Genet 1992;2:99–102.
31 Aguilera G, Pham Q, Rabadan-Diehl C: Regulation of pituitary vasopressin receptors during chronic stress: Relationship with corticotroph responsiveness. J Neuroendocrinol 1994;6:299–304.
32 Sugimoto T, Saito M, Mochizuki S, Watanabe Y, Hashimoto S, Kawashima H: Molecular cloning and functional expression of a cDNA encoding the human V1b vasopressin receptor. J Biol Chem 1994;269:27088–27092.
33 Lolait SJ, O'Carroll AM, Mahan LC, Felder CC, Button DC, Young WS III, Mezey E, Brownstein MJ: Extrapituitary expression of the rat V1b vasopressin receptor gene. Proc Natl Acad Sci USA 1995;92:6783–6787.
34 Raymond V, Leung PCK, Veilleux R, Labrie F: Vasopressin rapidly stimulates phosphatidic acid-phosphatidylinositol turnover in rat anterior pituitary cells. FEBS Lett 1985;182:196–200.
35 Todd K, Lightman SL: Vasopressin activation of phosphatidylinositol metabolism in rat anterior pituitary in vitro and its modification by changes in the hypothalamo-pituitary-adrenal axis. Neuroendocrinology 1987;45:212–218.
36 Bilezikjian LM, Blount AM, Vale WW: The cellular actions of vasopressin on corticotrophs of the anterior pituitary: Resistance to glucocorticoid action. Mol Endocrinol 1987;1:451–458.
37 Cronin MJ, Zysk JR, Baertschi AJ: Protein kinase C potentiates corticotrophin-releasing factor stimulated cyclic AMP in pituitary. Peptides 1986;7:935–938.
38 Abou-Samra AB, Harwood JP, Manganiello VC, Catt KJ, Aguilera G: Phorbol 12-myristate 13-acetate and vasopressin potentiate the effect of corticotrophin-releasing factor on cyclic AMP production in rat anterior pituitary cells. J Biol Chem 1987;262:1129–1136.

39 Al-Damluji S, Perry L, Tomlin S, Bouloux P, Grossman A, Rees LH, Besser GM: α-Adrenergic stimulation of corticotrophin secretion by a specific central mechanism in man. Neuroendocrinology 1987;45:68–76.

40 Takao T, Hashimoto K, Ota Z: Central catecholaminergic control of ACTH secretion. Regul Pept 1988;21:301–308.

41 Al-Damluji S: Review: Adrenergic mechanisms in the control of corticotrophin secretion. J Endocrinol 1988;119:5–14.

42 Itoi K, Suda T, Tozawa F, Dobashi I, Ohmori N, Sakai Y, Abe K, Demura H: Microinjection of norepinephrine into the paraventricular nucleus of the hypothalamus stimulates corticotrophin-releasing factor gene expression in conscious rats. Endocrinology 1994;135:2177–2182.

43 Makino S, Smith MA, Gold PW: Increased expression of corticotrophin-releasing hormone and vasopressin messenger RNA (mRNA) in the hypothalamic paraventricular nucleus during repeated stress: Association with reduction in glucocorticoid receptor mRNA levels. Endocrinology 1995; 136:3299–3309.

44 Suda T, Tozawa F, Iwai I, Sato Y, Sumitomo T, Nakano Y, Yamada M, Demura H: Neuropeptide-Y increases the corticotrophin-releasing factor messenger ribonucleic acid level in the rat hypothalamus. Brain Res Mol Brain Res 1993;18:311–315.

45 Kageyama K, Tozawa F, Horiba N, Watanobe H, Suda T: Serotonin stimulates corticotrophin-releasing factor gene expression in the hypothalamic paraventricular nucleus of conscious rats. Neurosci Lett 1998;243:17–20.

46 Balsa JA, Sanchez-Franco F, Pazos F, Lara JI, Lorenzo MJ, Maldonado G, Cacicedo L: Direct action of serotonin on prolactin, growth hormone, corticotrophin and luteinizing hormone release in cocultures of anterior and posterior pituitary lobes: Autocrine and/or paracrine action of vasoactive intestinal peptide. Neuroendocrinology 1998;68:326–333.

47 Jorgensen H, Knigge U, Kjær A, Warberg J: Adrenocorticotrophic hormone secretion in rats induced by stimulation with serotonergic compounds. J Neuroendocrinol 1999;11:283–290.

48 Ray DW, Ren SG, Melmed S: Leukaemia inhibitory factor regulates proopiomelanocortin gene transcription. Ann NY Acad Sci 1998;840:162–173.

49 Satoh H, Mori S: Subregional assignment of the proopiomelanocortin gene *(POMC)* to human chromosome band 2p23.3 by fluorescence in situ hybridization. Cytogenet Cell Genet 1997;76: 221–222.

50 Uhler M, Herbert E, D'Eustachio P, Roddle FD: The mouse genome contains two nonallelic proopiomelanocortin genes. J Biol Chem 1983;258:9444–9453.

51 Takahashi HY, Teranishi S, Nakanishi S, Numa S: Isolation and structural organization of the human corticotropin-β-lipotropin precursor gene. FEBS Lett 1981;135:97–102.

52 Cochet M, Chang AC, Cohen SN: Characterization of the structural gene and putative 5′-regulatory sequences for human proopiomelanocortin. Nature 1982;297:335–339.

53 Whitfeld PL, Seeburg PH, Shine J: The human pro-opiomelanocortin gene: Organization, sequence and interspersion with repetitive DNA. DNA 1982;1:133–143.

54 Jingami H, Nakanishi S, Imura H, Numa S: Tissue distribution of messenger RNAs coding for opioid peptide precursors and related RNA. Eur J Biochem 1984;142:441–447.

55 Jeannotte L, Burbach JPH, Drouin J: Unusual proopiomelanocortin ribonucleic acids in extrapituitary tissues: Intronless transcripts in testes and long poly(A) tails in hypothalamus. Mol Endocrinol 1987;1:749–757.

56 Lacaze-Masmonteil T, de Keyzer Y, Luton JP, Kahn A, Bertagna X: Characterization of proopiomelanocortin transcripts in human nonpituitary tissues. Proc Natl Acad Sci USA 1987;84:7261–7265.

57 DeBold CR, Nicholson WE, Orth DN: Immunoreactive proopiomelanocortin (POMC) peptides and POMC-like messenger ribonucleic acid are present in many rat nonpituitary tissues. Endocrinology 1988;122:2648–2657.

58 Chen CLC, Chang CC, Kreiger DT, Bardin CW: Expression and regulation of proopiomelanocortin-like gene in the ovary and placentas: Comparison with the testis. Endocrinology 1986;118: 2382–2389.

59 Chen CLC, Madigan MB: Regulation of testicular proopiomelanocortin gene expression. Endocrinology 1987;121:590–596.

60 Clark AJL, Lavender PM, Coates P, Johnson MR, Rees LH: In vitro and in vivo analysis of the peptide products of the short pro-opiomelanocortin mRNA. Mol Endocrinol 1990;4:1737–1743.
61 Murphy EP, Conneely OM: Neuroendocrine regulation of the hypothalamic pituitary adrenal axis by the nurr1/nur77 subfamily of nuclear receptors. Mol Endocrinol 1997;11:39–47.
62 Philips A, Lesage S, Gingras R, Maira MH, Gauthier Y, Hugo P, Drouin J: Novel dimeric Nur 77 signalling mechanism in endocrine and lymphoid cells. Mol Cell Biol 1997;17:5946–5951.
63 Maira M, Martens C, Philips A, Drouin J: Heterodimerization between members of the Nur subfamily of orphan nuclear receptors as a novel mechanism for gene activation. Mol Cell Biol 1999;19: 7549–7557.
64 Drouin J, Sun YL, Nemer M: Regulatory elements of the pro-opiomelanocortin gene: Pituitary specificity and glucocorticoid repression. Trends Endocrinol Metab 1990;219–225.
65 Philips A, Maira M, Mullick A, Chamberland M, Lesage S, Hugo P, Drouin J: Antagonism between Nur 77 and glucocorticoid receptor for control of transcription. Mol Cell Biol 1997;17:5952–5959.
66 Drouin J, Maira M, Philips A: Novel mechanism of action for Nur 77 and antagonism by glucocorticoids: A convergent mechanism for CRH activation and glucocorticoid repression of POMC gene transcription. J Steroid Biochem Mol Biol 1998;65:59–63.
67 Lamonerie T, Tremblay JJ, Lanctot C, Thierrien M, Gauthier Y, Drouin J: Ptx1, a bicoid-related homeobox transcription factor involved in transcription of the pro-opiomelanocortin gene. Genes Dev 1996;10:1284–1295.
68 Lanctot C, Gauthier Y, Drouin J: Pituitary homeobox 1 (Ptx 1) is differentially expressed during pituitary development. Endocrinology 1999;140:1416–1422.
69 Tremblay JJ, Drouin J: Egr-1 is a downstream effector of GnRH and synergizes by direct interaction with Ptx1 and SF-1 to enhance luteinizing hormone beta gene transcription. Mol Cell Biol 1998;19:2567–2576.
70 Tremblay JJ, Marcil A, Gauthier Y, Drouin J: Ptx1 regulates SF-1 activity by an interaction that mimics the role of the ligand binding domain. EMBO J 1999;18:3431–3441.
71 Tremblay JJ, Lanctot C, Drouin J: The pan-pituitary activator of transcription, Ptx1 (pituitary homeobox 1), acts in synergy with SF-1 and Pit1 and is an upstream regulator of the Lim-homeodomain gene Lim 3/Lhx3. Mol Endocrinol 1999;12:428–441.
72 Drouin J, Lamolet B, Lamonerie T, Lanctot C, Tremblay JJ: The PTX family of homeodomain transcription factors during pituitary developments. Mol Cell Endocrinol 1998;140:31–36.
73 Lanctot C, Lamolet B, Drouin J: The bicoid-related homeoprotein Ptx1 defines the most anterior domain of the embryo and differentiates posterior from anterior lateral mesoderm. Development 1997;124:2807–2817.
74 Logan M, Tabin CJ: Role of Pitx1 upstream of Tbx4 in specification of hindlimb identity. Science 1999;283:1736–1739.
75 Szeto DP, Rodriguez-Esteban C, Ryan AK, O'Connell SM, Liu F, Kioussi C, Gleiberman AS, Izpisua-Belmonte JC, Rosenfeld MG: Role of the bicoid-related homeodomain factor Pitx1 in specifying hindlimb morphogenesis and pituitary development. Genes Dev 1999;13:484–494.
76 Therrien M, Drouin J: Cell-specific helix-loop-helix factor required for pituitary expression of the pro-opiomelanocortin gene. Mol Cell Biol 1993;13:2342–2353.
77 Poulin G, Turgeon B, Drouin J: NeuroD1/β2 contributes to cell-specific transcription of the pro-opiomelanocortin gene. Mol Cell Biol 1997;17:6673–6682.
78 Malecki MT, Jhala US, Antonellis A, Fields L, Doria A, Orban T, Saad M, Warram JH, Montminy M, Krolewski AS: Mutations in NEUROD1 are associated with the development of type 2 diabetes mellitus. Nat Genet 1999;23:323–328.
79 Naya FJ, Huang HP, Qiu Y, Mutoh H, DeMayo FJ, Leiter AB, Tsai MJ: Diabetes, defective pancreatic morphogenesis, and abnormal enteroendocrine differentiation in β2/neuroD-deficient mice. Genes Dev 1997;11:2323–2334.
80 Roberts JL, Herbert E: Characterization of a common precursor to corticotropin and beta-lipotropin: Cell free synthesis of the precursor and identification of corticotropin peptides in the molecule. Proc Natl Acad Sci USA 1977;74:4826–4830.
81 Mains RE, Eipper BA, Ling N: Common precursor to corticotropins and endorphins. Proc Natl Acad Sci USA 1977;74:3014–3018.

82 Bertagna X, Nicholson WE, Sorenson GD, Pettengill OS, Mount CD, Orth DN: Corticotropin, lipotropin, and β-endorphin production by a human non-pituitary tumor in culture: Evidence for a common precursor. Proc Natl Acad Sci USA 1978;75:5160–5164.
83 Nakanishi S, Inoue A, Kita T, Nakamura M, Chang AC, Cohen SN, Numan S: Nucleotide sequence of cloned cDNA for bovine corticotropin-β-lipotropin precursor. Nature 1979;278:423–427.
84 Eipper BA, Mains RE: Structure and biosynthesis of pro-adrenocorticotropin-endorphin and related peptides. Endocr Rev 1980;1:1–27.
85 Cool DR, Normant E, Shen FS, Chen HC, Pannell L, Zhang Y, Loh YP: Carboxypeptidase E is a regulated secretory pathway sorting receptor: Genetic obliteration leads to endocrine disorders in *Cpe*^{fat} mice. Cell 1997;88:73–83.
86 Seidah NG, Chretien M: Complete amino acid sequence of a human pituitary glycopeptide: An important maturation product of pro-opiomelanocortin. Proc Natl Acad Sci USA 1981;78:4236–4240.
87 Smith AI, Funder JW: Proopiomelanocortin processing in the pituitary, central nervous system, and peripheral tissues. Endocr Rev 1988;9:159–179.
88 Zhou A, Bloomquist BT, Mains RE: The prohormone convertases PC1 and PC2 mediate distinct endoproteolytic cleavages in a strict temporal order during proopiomelanocortin biosynthetic processing. J Biol Chem 1993;268:1763–1769.
89 Fortenberry Y, Liu J, Lindberg I: The role of the 7B2 CT peptide in the inhibition of prohormone convertase 2 in endocrine cell lines. J Neurochem 1999;73:994–1003.
90 Westphal CH, Muller L, Zhou A, Zhu X, Bonner-Weir S, Schambelan M, Steiner DF, Lindberg I, Leder P: The neuroendocrine protein 7B2 is required for peptide hormone processing in vivo and provides a novel mechanism for pituitary Cushing's disease. Cell 1999;96:689–700.
91 Eipper BA, Bloomquist BT, Husten EJ, Milgram SL, Mains RE: Peptidylglycine α-amidating monooxygenase and other processing enzymes in the neurointermediate pituitary. Ann NY Acad Sci 1993;680:147–160.
92 Dahia PL, Grossman AB: The molecular pathogenesis of corticotroph tumors. Endocr Rev 1999;20:136–155.
93 Thodou E, Asa SL, Kontogeorgos G, Kovacs K, Horvath E, Ezzat S: Lymphocytic hypophysitis: Clinicopathological findings. J Clin Endocrinol Metab 1995;80:2302–2311.
94 Sauter NP, Toni R, McLoughlin CD, Dyess EM, Krizmann J, Lechan RM: Isolated ACTH deficiency associated with an autoantibody to corticotroph antigen that is not adrenocorticotrophin or other POMC-derived peptides. J Clin Endocrinol Metab 1990;70:1391–1397.
95 Muglia L, Jacobson L, Dikkes P, Majzoub JA: Corticotrophin-releasing hormone deficiency reveals major fetal but not adult glucocorticoid need. Nature 1995;373:427–432.
96 Venihaki M, Majzoub JA: Animal models of CRH deficiency. Front Neuroendocrinol 1999;20:122–145.
97 Weninger SC, Dunn AJ, Muglia LJ, Dikkes P, Miczek KA, Swiergiel AH, Berridge CW, Majzoub JA: Stress-induced behaviors require the corticotropin-releasing hormone (CRH) receptor, but not CRH. Proc Natl Acad Sci USA 1999;96:8283–8288.
98 Murch SH, Carter EP, Tsagarakis S, Grossman A, Savage MO: Isolated ACTH deficiency with absent response to corticotrophin-releasing factor-41: Evidence for a primary pituitary defect. Acta Paediatr Scand 1991 80:259–261.
99 Muglia LJ, Jacobson L, Weninger SC, Luedke CE, Bae DS, Jeong KH, Majzoub JA: Impaired diurnal adrenal rhythmicity restored by constant infusion of corticotropin-releasing hormone in corticotropin-releasing hormone-deficient mice. J Clin Invest 1997;99:2923–2929.
100 Koide Y, Kimura S, Inoue S, Ikeda M, Uchida K, Ando J, Shimizu A, Oda K, Itakura M, Nabeshima I, Maruyama M, Yamashita K, Abe K: Responsiveness of hypophyseal-adrenocortical axis to repetitive administration of synthetic ovine corticotrophin-releasing hormone in patients with isolated adrenocorticotrophin deficiency. J Clin Endocrinol Metab 1986;63:329–335.
101 Kyllo JH, Collins MM, Vetter KL, Cuttler L, Rosenfield RL, Donohoue PA: Linkage of congenital isolated adrenocorticotropic hormone deficiency to the corticotropin-releasing hormone locus using simple sequence repeat polymorphisms. Am J Med Genet 1996;62:262–267.

102 Timpl P, Spanagel R, Sillaber I, Kresse A, Reul JM, Stalla GK, Blanquet V, Steckler T, Holsboer F, Wurst W: Impaired stress response and reduced anxiety in mice lacking a functional corticotropin-releasing hormone receptor. Nat Genet 1998;19:162–166.
103 Turnbull AV, Smith GW, Lee S, Vale WW, Lee KF, Rivier C: CRF type I receptor-deficient mice exhibit a pronounced pituitary-adrenal response to local inflammation. Endocrinology 1999;140: 1013–1017.
104 Contarino A, Dellu F, Koob GF, Smith GW, Lee KF, Vale W, Gold LH: Reduced anxiety-like and cognitive performance in mice lacking the corticotropin-releasing factor receptor I. Brain Res 1999; 835:1–9.
105 Scaccianoce S, Muscolo LA, Cigliana G, Navarra D, Nicolai R, Angelucci L: Evidence for a specific role of vasopressin in sustaining pituitary-adrenocortical stress response in the rat. Endocrinology 1991;128:3138–3143.
106 Makara GB: The relative importance of hypothalamic neurons containing corticotropin-releasing factor or vasopressin in the regulation of adrenocorticotropic hormone secretion. Ciba Found Symp 1992;168:43–51.
107 Hansen LK, Rittig S, Robertson GL: Genetic basis of familial neurohypophyseal diabetes insipidus. Trends Endocrinol Metabol 1997;8:363–372.
108 Krude H, Biebermann H, Luck W, Horn R, Brabant G, Gruters A: Severe early-onset obesity, adrenal insufficiency and red hair pigmentation caused by POMC mutations in humans. Nat Genet 1998;19:155–157.
109 Huszar D, Lynch CA, Fairchild-Huntress V, Dunmore JH, Fang Q, Berkemeir LR, Gu W, Kesterson RA, Boston BA, Cone RD, Smith FJ, Campfield LA, Burn P, Lee F: Targeted disruption of the melanocortin 4 receptor results in obesity in mice. Cell 1997;88:131–141.
110 Yeo GSH, Farooqi S, Aminian S, Halsall DJ, Stanhope RG, O'Rahilly S: A frameshift mutation in *MC4R* associated with dominantly inherited human obesity. Nat Genet 1998;20:111–112.
111 Vaisse C, Clement K, Guy-Grand B, Froguel P: A frameshift mutation in human MC4R is associated with a dominant form of obesity. Nat Genet 1998;20:113–114.
112 Krude H, Gruters A: Implications of proopiomelanocortin (POMC) mutations in humans: The POMC deficiency syndrome. Trends Endocrinol Metabol 2000;11:15–22.
113 Hattori S, Onigata K, Takano Y, Harigaya A, Nagashima T, Morikawa A: Homozygous mutation for proopiomelanocortin gene causes congenital isolated adrenocorticotropin deficiency, early-onset obesity and less pigmented hair (abstract). 81st Annual Meeting of the Society for Endocrinology, 1999, P3–430.
114 Yaswen L, Diehl N, Brennan MB, Hochgeschwender U: Obesity in the mouse model of pro-opiomelanocortin deficiency responds to peripheral melanocortin. Nat Med 1999;5:1066–1070.
115 Naggert JK, Fricker LD, Varlamov O, Nishina PM, Rouille Y, Steiner DF, Carroll RJ, Paigen BJ, Leiter EH: Hyperproinsulinaemia in obese fat/fat mice associated with a carboxypeptidase E mutation which reduces enzyme activity. Nat Genet 1995;10:135–142.
116 O'Rahilly S, Gray H, Humphreys PJ, Krook A, Polonsky KS, White A, Gibson S, Taylor K, Carr C: Brief report: Impaired processing of prohormones associated with abnormalities of glucose homeostasis and adrenal function. N Engl J Med 1995;333:1386–1390.
117 Jackson RS, Creemers JW, Ohagi S, Raffin-Sanson ML, Sanders L, Montague CT, Hutton JC, O'Rahilly S: Obesity and impaired prohormone processing associated with mutations in the human prohormone convertase 1 gene. Nat Genet 1997;16:303–306.
118 Nussey SS, Soo SC, Gibson S, Gout I, White A, Bain M, Johnstone AP: Isolated congenital ACTH deficiency: A cleavage enzyme defect? Clin Endocrinol 1993;39:381–385.
119 Jackson RS, O'Rahilly S, Brain C, Nussey SS: Proopiomelanocortin products and human early-onset obesity. J Clin Endocrinol Metabol 1999;84:819–820.

Adrian J.L. Clark, Department of Endocrinology
St. Bartholomew's and the Royal London School of Medicine and Dentistry
West Smithfield, London EC1A 7BE (UK)
Tel. +44 20 7601 7445, Fax +44 20 7601 8468, E-Mail a.j.clark@mds.qmw.ac.uk

Rappaport R, Amselem S (eds): Hypothalamic-Pituitary Development. Genetic and Clinical Aspects.
Endocr Dev. Basel, Karger, 2001, vol 4, pp 162–174

Central Diabetes insipidus

Paul Czernichow

Endocrinology and Diabetes Unit, Hôpital Robert-Debré, Paris, France

Important insights into the aetiology of central diabetes insipidus (DI) in paediatrics have been provided by modern imaging techniques, particularly magnetic resonance imaging (MRI), which accurately visualizes the median eminence and pituitary stalk. Moreover, full characterization of the arginine-vasopressin-neurophysin II (AVP-NPII) gene has shed light on the mechanisms underlying familial DI due to structural abnormalities of this gene.

Aetiology

In infants, children and adolescents, causes of central or neurogenic DI include accidental or neurosurgical injury, tumours and other infiltrating lesions affecting the AVP-producing neurones, and structural AVP gene abnormalities. In three published series, 20–30% of cases were idiopathic [1–3]. Most of these cases were investigated by computed tomography (CT) of the hypothalamic-pituitary area. However, in a recent study of 70 cases [4], a similar proportion of cases remained idiopathic despite repeated MRI (table 1).

Any lesion of the pituitary stalk or median eminence interrupts the AVPergic axons that travel from the hypothalamus to the posterior pituitary lobe. This can result in retrograde degeneration of the hypothalamic neurones. Lesions closer to the hypothalamus affect more neurones, causing greater permanent loss of AVP secretion.

DI due to Neurosurgery or Head Injury

Neurosurgery: DI can develop after neurosurgery in patients whose preoperative AVP production was normal or partially deficient [5]. The symptoms usually

Table 1. Aetiology of hypothalamic DI in children and adults (% values) [adapted from 1–4]

	Children	Children + adults	Adults
Primary brain tumour	49.5	23	30
Before surgery	33.5	15.5	13
After surgery	16.0	7.5	17
Idiopathic (isolated or familial)	29	50[1]	25
Histiocytosis	16	15	
Metastatic cancer	–		8
Trauma	2.2	3	17
Post-infectious disease	2.2		–

[1] 27% had normal hypothalamopituitary structure by MRI.

set in within 1–4 days. Resolution over several days or weeks may occur. Less commonly, a triphasic pattern of polyuria with polydipsia, apparently normal urine concentration, and permanent polyuria occurs. The 2- to 14-day lag time between surgery and symptoms of DI is probably ascribable to AVP release from the disintegrating magnocellular neurones.

Head injury: The axons of AVP-containing magnocellular neurones extend without interruption to the posterior pituitary over a distance of approximately 10 mm. Trauma to the base of the brain can cause swelling around or severance of these axons, resulting in either transient or permanent DI. Permanent DI can occur after a seemingly minor injury. Approximately half of the patients with sella turcica fractures develop permanent DI [6], which may be delayed as long as 1 month after the injury, during which time the neurones or severed axons may undergo retrograde degeneration [7].

Tumours and Other Infiltrating Processes

The nuclei that synthesize AVP and regulate its secretion are bilateral; destruction of one side does not produce DI. However, immediately above the sellar diaphragm, the tracts from the two sides converge into the single supraopticohypophyseal tract. Lesions in this area cause significant impairment of AVP secretion. Tumours of the pituitary cause AVP deficiency only if they extend above the sella turcica.

Dysgerminoma is one of the tumour types most often responsible for DI. Dysgerminomas may secrete α-foetoprotein and/or β-hCG, which can cause precocious puberty in boys. Dysgerminoma cells may be found in the cerebrospinal fluid. The presence of these cells and/or elevated plasma or cerebrospinal spinal fluid concentrations of α-foetoprotein and hCG can help to determine the nature of a suprasellar mass associated with DI [8]. The tumour can remain too small to show up on MRI scans for several years after the onset of polyuria. For this reason, quantitative assays of the β subunit of hCG, which is often secreted by dysgerminomas and pinealomas, and MRI scans should be performed repeatedly in children with idiopathic DI. In some cases, plasma and cellular markers are absent, and the only manifestation of the disease is isolated thickening of the pituitary stalk [4, 9–11]. When serial MRIs demonstrate a fast-growing lesion, transsphenoidal biopsy may provide the diagnosis of dysgerminoma [4, 9].

Among patients with craniopharyngioma, 10–20% have severe DI, and approximately 35% have a partial AVP deficiency. However, the proportion of severe cases may be over-represented in reports since associated ACTH deficiency may mask the polyuria, leading to underdiagnosis of the less severe cases.

DI is also a common manifestation of histiocytosis X, causing infiltration of the posterior pituitary. The rate of occurrence of water balance disorders during the course of this disease varies from 10 to 30% [12–14]. DI can occur before, concurrently with, or after the development of lesions at extracranial sites. DI is more common in patients with skull lesions. Patients with extensive disease are more likely to have DI than those with lesions apparently confined to the bone [13]. Gadolinium-enhanced MRI may reveal thickening in the area of the hypothalamus and pituitary stalk [15].

Lymphocytic infundibulo-neurohypophysitis may be the cause of some cases of apparently idiopathic central DI [16]. This entity can occur in association with other autoimmune diseases [17]. Imaging studies disclose a large pituitary with a thick stalk [16, 18, 19]. Biopsy of the posterior pituitary in patients with this condition has shown lymphocytic infiltration of the hypothalamopituitary region. To our knowledge, this entity has never been described in children, the youngest patient described being 28 year old [16].

Whether antibody-mediated destruction of AVP-producing cells occurs is controversial. Antibodies directed against AVP-containing cells are found in more than half the patients with apparently idiopathic DI [20] and in some patients with other autoimmune diseases but no evidence of DI [21]. Indeed, in a large series of patients with endocrine autoimmune disease without evidence of central DI (878 patients), 9 patients were vasopressin cells antibody positive. Four of these cases showed impaired response to water deprivation test and during follow-up, 4 additional cases developed DI [22]. Thus, these antibodies may be mere markers for prior neurone destruction and DI even in patients without clinical evidence of vasopressin insufficiency.

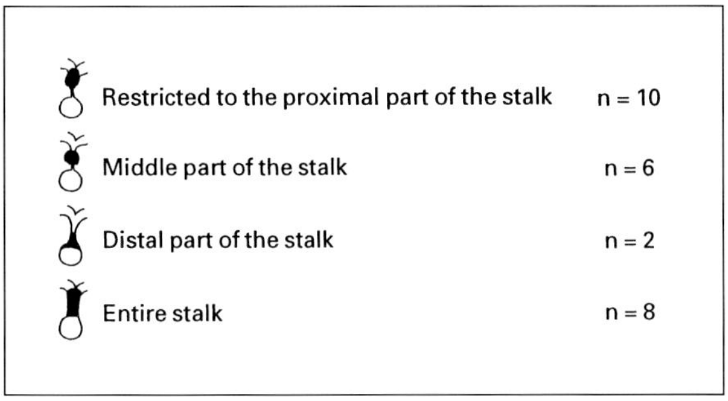

Fig. 1. Schematic representation of the pituitary stalk morphology in patients (n) with DI and pituitary stalk enlargement.

DI and Congenital Malformations of the Brain

Midline anatomic abnormalities of the brain such as septo-optic dysplasia with agenesis of the corpus callosum [23], Kabuki make-up syndrome [24], holoprosencephaly [25], and familial pituitary hypoplasia with absent stalk [26], are sometimes associated with central DI, even in the absence of external evidence of craniofacial abnormalities [25]. Central DI due to midline brain abnormalities is often accompanied with defects in thirst perception [24], suggesting that a common osmosensor may control both AVP release and thirst perception.

Thick Pituitary Stalk and Idiopathic DI

MRI allows a precise measure of the pituitary stalk. Its normal height varies from < 3 mm [4] to < 2 mm [10]. MRI has shown a thick pituitary stalk with or without thickening of the median eminence in some cases of idiopathic or secondary DI caused by infiltrating processes [10]. Figure 1 provides a schematic representation of this abnormality: the lesions usually involve either the entire stalk or the proximal part of the stalk, lesions of the distal stalk being less common.

The lesion may remain stable during several years. It may regress even to a complete reversal of pituitary stalk enlargement. In some cases a dramatic increase in size may be seen within a short time and seems typical of a dysgerminoma. However, the precise aetiology of DI with isolated pituitary stalk enlargement is known in a limited number of cases despite meticulous investigation and long-term follow-up. The most frequent lesions are histiocytosis and dysgerminoma. In a recent series of paediatric cases of DI and thick pituitary stalk, the lesion was due to a germinoma in 15% of the cases, to a histiocytosis X in 20% and remained idiopathic in 65% of the cases [10]. Similar percentages have been reported in children in another series of patients combining young adults [4].

Table 2. Pituitary stalk thickening and DI in 26 patients [adapted from 10]

Aetiology	Number of cases	
	at onset of DI	during follow-up (duration of follow-up)
Germinoma	0	4 (1.3 ± 0.6 years)
Histiocytosis	4	1 (4.4 years)
Idiopathic	22	17 (5.6 ± 3.6 years)

At onset of DI, 85% of the patients with pituitary stalk thickening had no proven aetiology. During follow-up, this percentage dropped to 65%. One patient was found to have histiocytosis X, and dysgerminoma was found in 4 patients.

In addition, the precise aetiology may no be apparent at onset of DI associated with stalk anomaly, which is the case for histiocytosis, but more importantly this is also the case for dysgerminoma [4, 10].

Dysgerminoma should be suspected in all patients with DI and pituitary stalk thickening, even when there are no neurological or ophthalmological symptoms. DI is usually the presenting symptom, although anterior pituitary dysfunction is a common concomitant. Most dysgerminomas grow rapidly, the mean time to the diagnosis of dysgerminoma was 1.3 years after the identification of pituitary stalk thickening by MRI and 2.5 years after the diagnosis of DI (table 2). MRI is therefore valuable for identifying small dysgerminomas.

Pituitary stalk biopsy has been suggested for identifying dysgerminoma in patients with DI and pituitary stalk thickening [9]. In one series of 9 patients with DI and pituitary stalk thickening, transsphenoidal biopsy of the stalk was performed in 6 cases and biopsy after craniotomy in another case and showed a dysgerminoma in each. However, it should be noted that in 1 case, two biopsies were performed: the first showed inflammatory cells and the second biopsy performed 2 years later was consistent with germinoma. We do not recommend pituitary stalk biopsy in patients with a thick but well-defined stalk <7 mm in diameter, as isolated idiopathic pituitary stalk thickening can resolve spontaneously (fig. 2). In patients with larger lesions, the presence of pituitary enlargement and anterior pituitary deficiencies should lead to a stronger suspicion of dysgerminoma and then a biopsy should be discussed. In most cases however, the cause of DI with isolated pituitary stalk thickening remains unknown [4, 10] and a long follow-up is necessary.

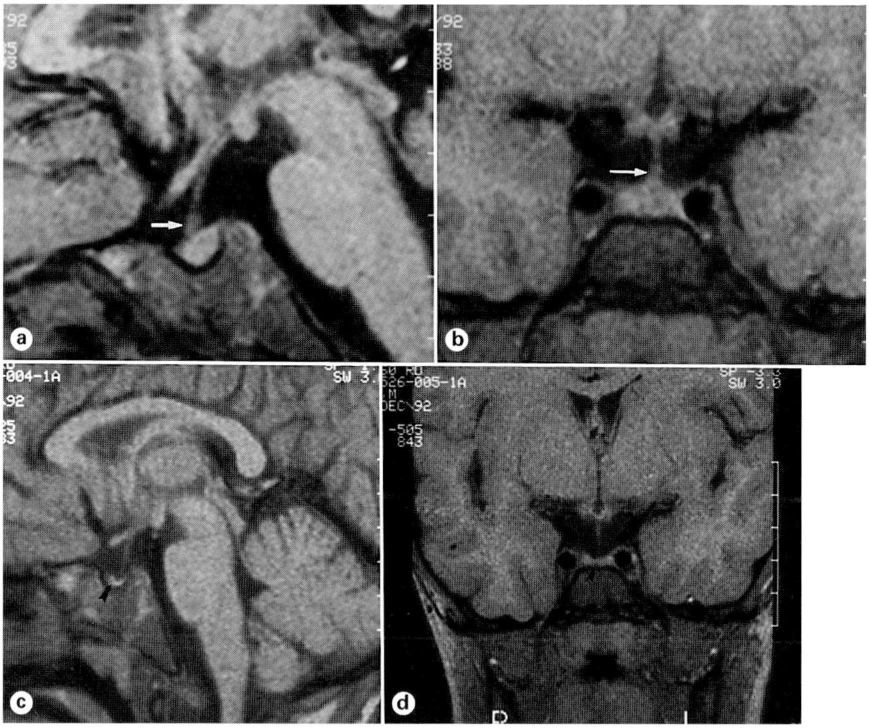

Fig. 2. MRI follow-up period of 1.9 years for a patient with idiopathic DI with isolated PST (patient 23). At the first evaluation (*a* sagittal slice, *b* coronal slice), the PS enlargement was restricted to the distal level of the PS (arrow). At the final evaluation 1.9 year later (*c* sagittal slice, *d* coronal slice), a decrease in anterior pituitary height (small portion; arrowhead) from 5 to 2.5 mm and a complete reversal of PS enlargement from 3 to 1.5 mm had occurred.

In conclusion, MRI is invaluable for investigating patients with DI. Several patients classified as having idiopathic DI without structural abnormalities of the hypothalamopituitary area have a pituitary stalk lesion. However, the exact nature of this lesion often remains unknown despite extensive and repeated investigations [27, 28].

Familial Central DI

Familial central DI is an autosomal dominant disorder of AVP secretion whose clinical manifestations resemble those of central DI due to other causes. Molecular genetic studies have clarified the basis of these cases.

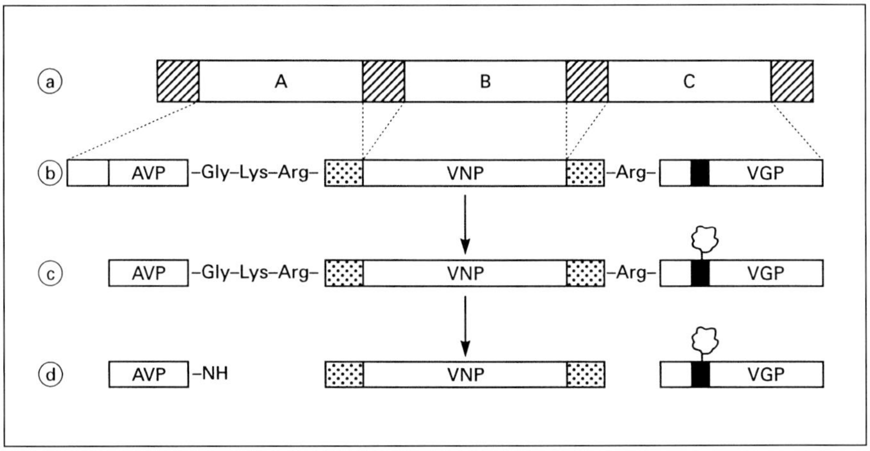

Fig. 3. A schematic representation of the AVP neurophysin gene (***a***) which codes for a preprohormone (***b***). It includes the signal peptide, the AVP peptide, the neurophysin (VNP) and a glycopeptide (VGP). The SP is cleaved away (***c***) and the three peptides packed in the granules (***d***).

The structure of the AVP-NPII gene is illustrated in figure 3. This gene is located on chromosome 20p3. The AVP-NPII gene contains three exons and produces a preprohormone that comprises the 19-amino-acid signal peptide, the 9-amino acid AVP peptide, a diamino acid linker, the 93-amino-acid NPII, a single amino acid linker, and, finally, the 39-amino-acid glycopeptide copeptin [29]. After synthesis of the preprohormone in the magnocellular neurone cell body, the signal peptide is cleaved away and the prohormone folds, placing AVP into a binding pocket of NPII. NPII is believed to protect AVP from proteolysis and to promote high-density packing in neurosecretory granules by oligomerization of AVP-NPII dimers. After formation of seven disulphide bonds within NPII and one within AVP and after glycosylation of copeptin, the prohormone is packaged into neurosecretory granules and cleaved into the product peptides during axonal transport to the posterior pituitary.

Patients with autosomal dominant DI have been found to be heterozygous for mutations in the AVP-NPII gene.

The mutations described to date are shown in table 3. Several signal peptide mutations have been identified [30–35] and shown to decrease the ability of the signal peptidase to initiate removal of the signal peptide from the preprohormone. A second group of mutations responsible for familial DI occurs within the AVP or amino-terminal domain of the NPII-coding sequence [33, 36–39]; these mutations modify the amino acids involved in the binding of AVP to NPII or in the

Table 3. Familial central DI and mutations of the AVP-NPII gene

Mutation	Amino acid change		Functional consequences
Coding sequence of signal peptide			
G → A at 279	Ala to Thr at SPI	}	Inefficient processing of prepropeptide [30–35]
C → T at 280	Ala to Val at SP-1		
C → T at 274	Ser to Phe at SP-3		
Coding sequence of neurophysin			
G → T at 1740	Gly to Val at amino acid 17	}	Alteration in conformation of AVP-NP II precursor [36–41]
Deletion of AGG	Deletion of Glu at amino acid 47		
G → T at 1859	Gly to Ser at amino acid 57		
G → T at 1874	Gly to Trp at amino acid 62		
C → T at 1761	Pro to Leu at amino acid 24		
G → T at 1758	Gly to Val at amino acid 23		
C → A at 1891	Cys to Stop at amino acid 67	→	Truncated neurophysin [40]
Coding sequence of vasopressin			
C → T at 301	Leu to PRO	→	Production of a weak agonist of vasopressin

folding of NPII. Finally, a third group of mutations results in synthesis of a truncated neurophysin molecule [40].

In most cases of familial DI, the symptoms develop during the first 10 years of life. Neonatal manifestations are uncommon, suggesting that one or both AVP-NPII genes initially ensure production of sufficient AVP to maintain an adequate water balance. A report of a gradual decline in plasma AVP levels during a 3-year period in a child with familial DI is in favour of this hypothesis [31]. Finally, autopsy findings of magnocellular neurone degeneration in symptomatic individuals [41, 42] suggest that the pathophysiology of familial central DI involves progressive postnatal degeneration of AVP-producing magnocellular neurones.

The possibility that a mutation of a gene encoding a protein secreted by neurones may be responsible for progressive degeneration of AVP-producing cells has been investigated in recent years. It has been postulated that all the mutations described to date may be important for proper folding and processing of the preprohormone [35]. As a result of this abnormality, the mutant protein may accumulate within the endoplasmic reticulum, where it may kill the cells by interfering with the orderly processing of other essential proteins. This hypothesis is supported by recent in vitro expression studies demonstrating accumulation of mutant AVP-NPII precursor within the endoplasmic reticulum [35].

These findings, together with the clinical picture of familial DI, suggest that almost all the mutations described to date result in production of an abnormal preprohormone that cannot be processed properly and eventually destroys the AVP-processing neurones. Perturbations in protein folding and assembly which result in abnormal function and are responsible for the pathological changes in one organ are emerging as an important mechanism of disease. Several genetic disorders are described in which protein misfolding has a major role [43].

As most cases reported so far have an autosomal dominant disorder due to a mutation affecting the synthesis of signal peptide of neurophysin, a single family with an autosomal recessive DI has also been reported. All 3 affected children were homozygous and the parents heterozygous for a single novel mutation in exon I replacing PRO of mature AVP with Leu (Leu-AVP) [44]. This peptide is a weak agonist of AVP with a reduced binding of the peptide to V2 receptor explaining the polyuria.

Wolfram syndrome is an autosomal recessive disorder and has been linked to markers on the short arm of chromosome 4. The syndrome includes: DI, diabetes mellitus, optic atrophy and sensorineural deafness. The DI is of gradual onset and usually partial. An autosomal deletion seems to be responsible for the occurrence of mitochondrial deletion [45].

Clinical Presentation

Usual Clinical Manifestations

In DI starting after infancy, the presenting symptoms are excessive urine output and thirst, usually with an abrupt onset. Most patients prefer water to other fluids. Polyuria may cause enuresis. Nocturia and the need to get up to drink several times during the night disturbs sleep, often causing tiredness responsible for poor academic achievement. Overt hypertonic dehydration occurs only if for some reason the patient is unable to obtain water. Food consumption is often markedly decreased, and this can cause loss of weight and slowing of growth.

Infantile DI

In neonates and infants, the clinical picture of DI is quite different from that in older patients and is difficult to recognise. Yet, early diagnosis is of the utmost importance, as in infants too young to express a need for water the risk of brain damage is high. The common leading manifestations are repeated episodes of hyperthermia and vomiting. Failure to thrive and constipation may be the only presenting symptoms.

Diagnosis

The diagnosis relies on measurements of plasma and urine osmolality and on the desmopressin test [46]. Urine osmolality normally shows a defined relationship to plasma osmolality. Since this relationship is mediated by plasma AVP, in most cases of DI, measuring AVP adds little to the information obtained by measuring osmolality. In nephrogenic DI, the urine-plasma osmolality relationship is disrupted, and plasma AVP measurement is useful as it shows abnormally high levels in comparison to plasma osmolality. The work-up in patients with DI should include determinations of serum creatinine or urea, sodium, potassium, calcium and glucose. Other investigations should be done to look for a cause, and as discussed above the key procedure is MRI of the hypothalamopituitary region.

Treatment

The goal of therapy is to maintain a normal water balance allowing the patient to lead a normal life without nocturia and with a normal daytime urine output. This goal was not fully achievable with the preparations available before desmopressin was introduced. Chlorpropamide, carbamazepine and clofibrate have no role in the treatment of paediatric DI.

Desmopressin (deamino-*D*-arginine AVP, DDAVP) is a synthetic AVP analogue with a prolonged antidiuretic effect and virtually no vasopressor activity. It is highly satisfactory in paediatric DI. A nasal spray is available (desmopressin acetate, 0.1 mg/ml). Five to 20 μg administered twice daily is usually adequate. Small premature infants have been treated with approximately 1 μg/kg daily, divided into two doses [47]. For such small amounts, the nasal solution is diluted 1:10 in 0.9% NaCl solution. The best approach is to start with a low dose, e.g., 1 μg for infants and 2.5 μg for young children, at 12-hour intervals, and to titrate the dose upwards until a satisfactory effect is achieved. During episodes of rhinitis, the dosage requirement may increase slightly, even if the nose is very carefully cleaned before administration of the spray. Headaches, nasal congestion and abdominal discomfort are rare side effects. Hyponatraemia should be looked for periodically because some patients tend to take excessive dosages to reduce their urine output.

Recently, oral desmopressin has been described as an attractive alternative for children [48–50]. The required dosage is 20 times larger than that of the nasal spray nasal dosage, which may make the oral drug more expensive. However, oral administration may be useful during periods of rhinitis and in disabled patients.

The patient must have free access to water. However, family members and carers must be aware that replacement therapy makes the patient unable to eliminate excess water, for instance during a drinking contest. They should also be informed that if the patient is also on cortisol replacement therapy, discontinuation of the cortisol carries a risk of hyponatraemia.

In patients who develop DI immediately after surgery, some physicians prefer aqueous AVP because of its shorter action [48]. Parenteral desmopressin solution (desmopressin acetate 4 μg/ml) is a good alternative. The dosage ranges from 0.5 to 2 μg/m^2 and should be tailored to the needs of each individual patient.

References

1 Czernichow P, Pomarede R, Brauner R, Rappaport R: Neurogenic diabetes insipidus in children; in Czernichow P, Robinson AG (eds): Diabetes insipidus in Man. Basel, Karger, 1985, pp 190–209.

2 Greger NG, Kirkland RT, Clayton GW, Kirkland JL: Central diabetes insipidus. Am J Dis Child 1986;140:551–554.

3 Wang LC, Cohen ME, Duffner PK: Etiologies of central diabetes insipidus in children. Pediatr Neurol 1994;11:273–277.

4 Maghnie M, Cosi G, Genoves E, Manca-Bitti ML, Cohen A, Zecca S, Tinelli C, Galluci M, Bernasconi S, Boscherini B, Severi F, Arico M: Central diabetes insipidus in children and young adults. N Engl J Med 2000;343:998–1007.

5 Seckl JR, Dunger DB, Lightman SL: Neurohypophyseal peptide function during early postoperative diabetes insipidus. Brain 1987;110:737.

6 Defoer F, Mahler C, Dua G et al: Posttraumatic diabetes insipidus. Acta Anaesthesiol Belg 1987;38: 397–399.

7 Hadani M, Findler G, Shaked I et al: Unusual delayed onset of diabetes insipidus following closed head trauma: Case report. J Neurosurg 1985;63:456–458.

8 Pomarede R, Czernichow P, Finidori J et al: Endocrine aspects of tumoral markers in intracranial germinomas. J Pediatr 1982;101:374–378.

9 Mootha SI, Barkovitch AJ, Grumbach MH, Edwards MS, Gitelman SE, Kaplan SL, Conte FA: Idiopathic hypothalamic diabetes insipidus, pituitary stalk thickening and the occult intracranial germinoma in children and adolescents. J Clin Endocrinol Metab 1997;82:1362–1367.

10 Léger J, Velasquez A, Garel C, Hassan M, Czernichow P: Thickened pituitary stalk on magnetic resonance imaging in children with central diabetes insipidus. J Clin Endocrinol Metab 1999;84: 1954–1960.

11 Maghnie M, Bossi G, Klersy C, Cosi G, Genovese E, Arico M: Dynamic endocrine testing and magnetic resonance imaging in the long-term follow-up of childhood Langerhans cell histiocytosis. J Clin Endocrinol Metab 1998;83:3089–3094.

12 Grois N, Flucher-Wolfram B, Heitger A, Mostbeck GH, Hofmann J, Gadner H: Diabetes insipidus in Langerhans' cell histiocytosis: Results from the DAL-HX 83 study. Med Pediatr Oncol 1995;24: 248–256.

13 Dunger DB, Broadbent V, Yeoman E, Sekl JR, Lightman SL, Grant DB, Pritchard J: The frequency and natural history of diabetes insipidus in children with Langerhans' cell histiocytosis. N Engl J Med 1989;321:1157–1162.

14 The French Langerhans' Cell Histiocytosis Study Group: A multicenter retrospective survey of Langerhans' cell histiocytosis: 343 cases observed between 1983 and 1993. Arch Dis Child 1996;75: 17–24.

15 Rosenfield DS, Abrahams J, Komp D: Brain MR in patients with Langerhans' cell histiocytosis: Findings and enhancement with Gd-DTPA. Pediatr Radiol 1990;20:433–436.

16 Imura H, Nakao K, Shimatsu A et al: Lymphocytic infundibuloneurohypophysitis as a cause of central diabetes insipidus. N Engl J Med 1993;329:683–689.
17 Paja M, Estrada J, Ojeda A et al: Lymphocytic hypophysitis causing hypopituitarism and diabetes insipidus, and associated with autoimmune thyroiditis, in a non-pregnant woman. Postgrad Med J 1994;70:220–224.
18 Koshiyama H, Sato H, Yorita S et al: Lymphocytic hypophysitis presenting with diabetes insipidus: Case report and literature review. Endocr J 1994;41:93–97.
19 Kojima H, Nojima T, Nagashima K et al: Diabetes insipidus caused by lymphocytic infundibuloneurohypophysitis. Arch Pathol Lab Med 1989;113:1399–1401.
20 Scherbaum WA, Czernichow P, Bottazzo GF, Doniach D: Diabetes insipidus in children. A possible autoimmune type with vasopressin cell antibodies. J Pediatr 1985;107:922–925.
21 De Bellis A, Bizzarro A, Amoresano Paglionico V, Di Marino S, Cziscudo T, Sinisi AA, Lombardi G, Bellastella A: Detection of vasopressin cell antibodies in some patients with autoimmune endocrine diseases without overt diabetes insipidus. Clin Endocrinol 1994;40:173–177.
22 De Bellis A, Colao A, Di Salle F, Muccitelli VI, Iorio S, Perrino S, Pivonello R, Coronella C, Bizzarro A, Lombardi G, Bellastella A: A longitudinal study of vasopressin cell antibodies, posterior pituitary function and magnetic resonance imaging evaluations in subclinical autoimmune central diabetes insipidus. J Clin Endocrinol Metab 1999;84:3047–3051.
23 Masera N, Grant DB, Stanhope R, Preece MA: Diabetes insipidus with impaired osmotic regulation in septo-optic dysplasia and agenesis of the corpus callosum. Arch Dis Child 1994;70:51–53.
24 Tawa R, Kaino Y, Ito T et al: A case of Kabuki make-up syndrome with central diabetes insipidus and growth hormone neurosecretory dysfunction. Acta Paediatr Jpn 1994;36:412.
25 Van Gool S, De Zegher F, De Vries LS et al: Alobar holoprosencephaly, diabetes insipidus and coloboma without craniofacial abnormalities: A case report. Eur J Pediatr 1990;149:621–622.
26 Yagi H, Nagashima K, Miyake H et al: Familial congenital hypopituitarism with central diabetes insipidus. J Clin Endocrinol Metab 1994;78:884–889.
27 Telucksingh S, Sellar R, Seckl JR, Edwards CRW, Padfield PL: Reversible pituitary stalk enlargement in cranial diabetes insipidus. J Neurol Neurosurg Psychiatry 1991;10:937–938.
28 Sato N, Sze G, Endo K: Hypophysitis: Endocrinologic and dynamic MR findings. Am J Neuroradiol 1998;19:439–444.
29 Miller XL: Molecular genetics of familial central diabetes insipidus. J Clin Endocrinol Metab 1993;77:592–595.
30 Ito M, Oiso Y, Murase T, Kondo K, Saito H, Chenzei T, Racchi M et al: Possible involvement of inefficient cleavage of preprovasopressin by signal peptidase as a cause for familial central diabetes insipidus. J Clin Invest 1993;91:2565–2571.
31 McLeod J, Kovacs L, Gaskill M, Rittig S, Bradley G, Robertson G: Familial neurohypophyseal diabetes insipidus associated with a signal peptide mutation. J Clin Endocrinol Metab 1993;77:599A–599G.
32 Krishnamani MR, Phillips JA, Copeland KC: Detection of a novel arginine vasopressin defect by dideoxy fingerprinting. J Clin Endocrinol Metab 1993;77:596–598.
33 Rittig S, Robertson GL, Siggaard C, Kovacs L, Gregersen N, Nyborg J, Pedersen EB: Identification of 13 new mutations in the vasopressin-neurophysin II gene in 17 kindreds with familial autosomal dominant neurohypophyseal diabetes insipidus. Am J Hum Genet 1996;58:107–117.
34 Repaske DR, Medlej R, Gültekin EK, Krishnamani MRS, Halaby G, Findling JW, Phillips JA III: Heterogeneity in clinical manifestation of autosomal dominant neurohypophyseal diabetes insipidus caused by mutation encoding $Ala^{-1} \rightarrow Val$ in the signal peptide of the arginine vasopressin/neurophysin II/copeptin precursor. J Clin Endocrinol Metab 1997;82:51–56.
35 Siggaard C, Rittig S, Corydon TJ, Andreasen PH, Jensen TG, Andresen BS, Robertson GL, Gregersen N, Bolund L, Pedersen EB: Clinical and molecular evidence of abnormal processing and trafficking of the vasopressin preprohormone in a large kindred with familial neurohypophyseal diabetes insipidus due to a signal peptide mutation. J Clin Endocrinol Metab 1999;84:2933–2941.
36 Bahnsen U, Oosting P, Swaab DF, Nahke P, Richter D, Schmale H: A missense mutation in the vasopressin-neurophysin precursor gene cosegregates with human autosomal dominant neurohypophyseal diabetes insipidus. EMBO J 1992;11:19–29.

37 Repaske DR, Browning JE: A de novo mutation in the coding sequence for neurophysin-II (Pro[24] → Leu) is associated with onset neurohypophyeal diabetes insipidus. J Clin Endocrinol Metab 1994; 74:421–427.
38 Gagliardi PC, Bernasconi S, Repaske DR: Autosomal dominant neurohypophyseal diabetes insipidus associated with missense mutation encoding Gly[23] → Val in neurophysin. J Clin Endocrinol Metab 1997;82:3643–3646.
39 Ito M, Mori Y, Oiso Y, Saito H: A single base substitution in the coding region for neurophysin II associated with familial central diabetes insipidus. J Clin Invest 1991;87:725–728.
40 Nagasaki H, Ito M, Yuasa H, Saito H, Fukase M, Hamada K, Ishikawa E, Katakami H, Oiso Y: Two novel mutations in the coding region for neurophysin II associated with familial central diabetes insipidus. J Clin Endocrinol Metab 1995;80:1352–1356.
41 Braverman LE, Mancini JP, McGoldrich DM: Hereditary idiopathic diabetes insipidus. A case report with autopsy findings. Ann Intern Med 1965;63:503–508.
42 Bergeron C, Kovacs K, Ezrin C, Mizzen C: Hereditary diabetes insipidus: An immunohistochemical study of the hypothalamus and pituitary gland. Acta Neuropathol 1991;81:345–348.
43 Kuznetsov G, Nigam SK: Folding of secretory and membrane proteins. N Engl J Med 1998;339: 1688–1695.
44 Willcutts MD, Felner E, White PC: Autosomal recessive familial neurohypophyseal diabetes insipidus with continued secretion of mutant weakly active vasopressin. Hum Mol Genet 1999;8:1303–1307.
45 Hofmann S, Bezold R, Jaksch M, Kaufhold P, Obermaier-Kusser B, Gerbitz KD: Analysis of the mitochondrial DNA from patients with Wolfram (DIDMOAD) syndrome. Mol Cell Biochem 1997; 174:209–213.
46 Czernichow P: Testing water regulation; in Rank MB (ed): Diagnostic of Endocrine Function in Children and Adolescents, ed 2. Heidelberg, Barth, 1996, pp 230–240.
47 Giacoia GP, Watson S, Karathanos A: Treatment of neonatal diabetes insipidus with desmopressin. South Med J 1984;77:75–81.
48 Fjellestad-Paulsen A, Tubiana-Rufi N, Harris A, Czernichow P: Central diabetes insipidus in children. Antidiuretic effect and pharmacokinetics of intranasal and peroral 1-deamino-8-*D*-arginine vasopressin. Acta Endocrinol (Copenh) 1987;115:307–312.
49 Fjellestad-Paulsen A, Czernichow P: Central diabetes insipidus in children. V. Oral treatment with a vasopressin analogue (DDAVP). Acta Paediatr Scand 1986;75:605–610.
50 McDonald JA, Martha PM, Kerrigan J: Treatment of the young child with postoperative central diabetes insipidus. Am J Dis Child 1989;143:201–206.

Dr. Paul Czernichow, Endocrinology and Diabetes Unit
Hôpital Robert-Debré, 48, blvd Sérurier, F–75019 Paris (France)
Tel. +33 1 40 03 23 60, E-Mail paul.czernichow@rdb.ap-hop-paris.fr

Author Index

Subject Index